Chapter 1: Introduction to Dopamine – The Brain's Reward System

Dopamine is a crucial neurotransmitter in the brain, influencing everything from our sense of pleasure to our ability to move and think clearly. Often referred to as the brain's "reward" chemical, dopamine is essential to the way we experience motivation, happiness, and even addiction. But while dopamine's role in pleasure and reward is widely known, the full scope of its functions in regulating our behavior and cognition is far more intricate and multifaceted.

In this chapter, we will explore the foundational role of dopamine in the brain and introduce the two key dopamine receptors that are central to this book—**DRD1** and **DRD2**. Understanding these receptors' roles is essential for mastering dopamine's impact on both our emotional and cognitive lives.

The Role of Dopamine in the Brain

Dopamine is often described as a "feel-good" neurotransmitter because it plays a pivotal role in our experience of pleasure and satisfaction. However, dopamine's functions extend far beyond simple reward. It is involved in regulating mood, attention, learning, decision-making, and even motor control.

The brain releases dopamine in response to rewarding stimuli, reinforcing behaviors that increase the likelihood of their repetition. When we achieve something we value—whether it's eating a favorite food, receiving praise, or achieving a goal—dopamine is released, creating a feeling of pleasure or satisfaction. This dopamine surge serves to reinforce the behavior, encouraging us to repeat it in the future. It's no wonder that dopamine is often associated with motivation—it's the drive that pushes us toward our goals.

But dopamine's role in the brain is not just about the experience of pleasure. It also helps regulate arousal, attention, and the brain's response to stress. Additionally, dopamine plays a central role in learning and memory, particularly in how we anticipate and prepare for rewards. This complex interplay is what makes dopamine a key player in virtually every aspect of human experience—from how we think, learn, and perform to how we emotionally respond to the world around us.

Dopamine Receptors – The Key to Dopamine's Influence

Dopamine does not act alone. It exerts its effects through specific receptors located throughout the brain. These receptors are proteins on the surface of neurons that receive and process dopamine signals. The two main types of dopamine receptors—**D1-like receptors (DRD1)** and **D2-like receptors (DRD2)**—each have distinct functions but work together to regulate a broad array of brain activities.

1. **D1 Receptors (DRD1)**:

    The DRD1 receptors are primarily involved in modulating the brain's reward system and cognitive functions like attention, learning, and motivation. When dopamine binds to these receptors, it influences areas of the brain involved in memory, goal-setting, and decision-making. DRD1 receptors are highly active in areas like the prefrontal cortex and hippocampus, which are crucial for executive functions and memory consolidation. Activation of these receptors can enhance the brain's flexibility, helping us to learn new things, adapt to changing circumstances, and pursue rewards effectively.

2. **D2 Receptors (DRD2)**:

    The DRD2 receptors, on the other hand, are involved in regulating pleasure, habit formation, and motor control. These receptors play a key role in how we experience reinforcement and how habits are formed. DRD2 receptors are closely associated with the brain's basal ganglia, which are responsible for controlling movement and habit-related learning. DRD2 dysfunction has been implicated in a variety of conditions, including addiction, depression, and movement disorders like Parkinson's disease.

While DRD1 and DRD2 perform distinct roles, they are not isolated in their functions. Instead, they interact with one another in complex ways that regulate the balance between motivation, reward, and behavior. The dynamic interplay between these two receptors is critical in shaping our thoughts, actions, and emotional responses.

Introduction to DRD1 and DRD2: Why They Matter

The DRD1 and DRD2 receptors are not just important for understanding the mechanics of the brain's reward system—they are also central to understanding mental health, addiction, and even cognitive performance. By mastering how these receptors work, we can optimize our approach to everything from motivation and goal-setting to emotional regulation and behavior change.

- **Motivation and Goal Achievement**: Understanding how DRD1 influences motivation and how DRD2 regulates habit formation can provide insight into how we pursue goals. When dopamine activates DRD1 receptors, it strengthens our drive to accomplish objectives, while DRD2 plays a role in determining how habits and behaviors become ingrained in our daily lives.
- **Emotional and Cognitive Balance**: A well-balanced dopamine system, with proper function of both DRD1 and DRD2, is crucial for emotional regulation and cognitive clarity. Imbalances in dopamine signaling can lead to mood disorders, cognitive impairments, and even addictions. Thus, understanding these receptors provides an opportunity to optimize mental health.
- **Addiction and Reward**: Dopamine's involvement in addiction is another reason why DRD1 and DRD2 are so important. Overactivation of these receptors, particularly DRD2, can lead to compulsive behavior and reward-seeking that can escalate into addiction. Conversely, low levels of DRD2 activation are associated with a reduced ability to experience pleasure and motivation, which can contribute to conditions like depression.

Dopamine Receptors and Personal Growth

In the chapters to follow, we will explore the molecular and physiological workings of these dopamine receptors in greater detail. We will examine the science behind dopamine synthesis, the pathways it travels in the brain, and the critical role it plays in addiction, motivation, and cognitive function. By understanding the mechanisms at play, we will learn how to tap into the power of dopamine for a more balanced and fulfilling life.

In the next chapter, we will delve into the science of dopamine itself—how it is synthesized, released, and reabsorbed by the brain. We will also look at how the brain's reward pathways influence our daily behavior and the way we experience the world.

With this foundation of understanding, you are now prepared to unlock the potential of dopamine and its receptors, gaining insight into how mastering them can lead to greater cognitive clarity, emotional balance, and overall well-being.

Chapter 2: The Science of Dopamine – How It Works

Dopamine is one of the most important neurotransmitters in the brain, influencing a vast array of cognitive, emotional, and physical functions. But how does dopamine exert its powerful influence? In this chapter, we will explore the fundamental science behind dopamine's synthesis, release, and reuptake, as well as its pathways in the brain and the wide-ranging physiological and psychological effects it produces.

Dopamine Synthesis – From Precursors to Activation

Dopamine is synthesized from the amino acid **tyrosine**, which is derived from dietary protein. The process begins when **tyrosine hydroxylase**, the rate-limiting enzyme, converts tyrosine into **L-DOPA**, a precursor of dopamine. L-DOPA is then decarboxylated into dopamine by the enzyme **DOPA decarboxylase**. This process primarily occurs in dopaminergic neurons, particularly in regions like the substantia nigra, the ventral tegmental area (VTA), and the hypothalamus, which are key areas involved in reward, movement, and regulation of mood.

Dopamine is produced in a variety of regions of the brain, each of which has distinct functions. The **substantia nigra** and **striatum** are primarily responsible for motor control, while the **ventral tegmental area (VTA)** is central to the brain's reward pathways. Understanding where dopamine is produced and how it interacts with different regions of the brain is key to appreciating its wide-ranging effects on behavior and mental states.

Dopamine Release – The Brain's Reward Signal

Once synthesized, dopamine is stored in synaptic vesicles within neurons. When an electrical signal, known as an **action potential**, reaches the neuron's axon terminal, dopamine is released into the synaptic cleft (the gap between two neurons). This release is often triggered by stimuli that the brain perceives as rewarding—whether it's the anticipation of a goal, the experience of a pleasurable sensation, or the fulfillment of a need. The release of dopamine is a key component of the brain's reward system, which reinforces behaviors that lead to positive outcomes.

The dopamine released into the synaptic cleft then binds to dopamine receptors on the postsynaptic neuron. As mentioned earlier, the two most important dopamine receptors are **DRD1** and **DRD2**, each of which plays a role in different aspects of brain function and behavior.

When dopamine binds to DRD1 receptors, it activates a cascade of signaling pathways that enhance neural activity, particularly in areas related to motivation, learning, and reward. Conversely, binding to DRD2 receptors leads to a different set of signals that regulate pleasure-seeking behavior, motor control, and habit formation.

Dopamine Reuptake – Maintaining Balance

After dopamine is released and has carried out its function, it must be cleared from the synaptic cleft to prevent overstimulation. This is achieved through **dopamine reuptake**, a process where dopamine is reabsorbed into the presynaptic neuron by the **dopamine transporter (DAT)**. The dopamine that is taken back into the neuron can be repackaged into synaptic vesicles for future use or broken down by enzymes such as **monoamine oxidase (MAO)** and **catechol-O-methyltransferase (COMT)**.

Reuptake is an essential part of maintaining the delicate balance of dopamine in the brain. If dopamine is not reabsorbed efficiently, it can lead to excessive stimulation of dopamine receptors, which may result in conditions like addiction or hyperactivity. Conversely, if dopamine is reuptaken too quickly or not released in sufficient quantities, it can lead to mood disorders, such as depression or anhedonia (the inability to experience pleasure).

Dopamine Pathways – Mapping the Brain's Reward System

The brain has several distinct dopamine pathways, each associated with different aspects of behavior, motivation, and cognition. These pathways include:

1. **Mesolimbic Pathway (Reward Pathway)**:

    The mesolimbic pathway is one of the most well-known dopamine circuits, responsible for reward processing and pleasure. It originates in the VTA (ventral tegmental area) and projects to areas like the **nucleus accumbens**, which is often referred to as the brain's "pleasure center." The mesolimbic pathway is activated when you engage in activities that bring pleasure, such as eating, socializing, or achieving a goal. It is also implicated in addiction, as drugs and behaviors that stimulate this pathway can reinforce compulsive behaviors.

2. **Mesocortical Pathway (Cognition and Executive Function)**:

    This pathway also originates in the VTA but projects to the **prefrontal cortex**, which is involved in higher cognitive functions like decision-making, planning, and impulse control. Dopamine in this pathway helps maintain attention, working memory, and cognitive flexibility. Dysregulation of this pathway is linked to cognitive disorders like schizophrenia, where problems with executive function and attention are common.

3. **Nigrostriatal Pathway (Movement and Habit Formation):**

   The nigrostriatal pathway begins in the **substantia nigra** and projects to the **striatum**, which is involved in motor control and the formation of habits. Dopamine in this pathway helps coordinate movement, and its dysfunction is a hallmark of **Parkinson's disease**, a neurodegenerative disorder characterized by tremors, rigidity, and bradykinesia (slowness of movement).

4. **Tuberoinfundibular Pathway (Regulation of Hormones):**

   This pathway connects the hypothalamus to the pituitary gland and regulates the release of hormones, including **prolactin**, which is involved in lactation. Dopamine acts as an inhibitory signal to prevent the overproduction of prolactin.

The balance between these pathways and the efficiency of dopamine release and receptor activity in each of them have profound effects on mood, behavior, cognition, and physical performance.

Physiological and Psychological Effects of Dopamine

Dopamine's effects on the brain and body are broad and profound. On a physiological level, dopamine is directly involved in controlling motor function, rewarding pleasurable activities, and regulating the release of certain hormones. On a psychological level, dopamine influences motivation, mood, attention, and the ability to learn.

1. **Motivation and Goal-Directed Behavior:**

    Dopamine is often referred to as the "motivation molecule" because it drives us to pursue rewards. Dopamine release increases when we anticipate a reward, and this anticipatory response plays a critical role in how we pursue goals. The dopamine system encourages goal-directed behavior by reinforcing actions that lead to positive outcomes, thus creating a feedback loop that strengthens motivated behavior.

2. **Mood and Emotional Regulation:**

    Dopamine also plays a significant role in regulating mood and emotional well-being. Low dopamine levels are associated with depression, anhedonia (lack of pleasure), and fatigue, while elevated dopamine can lead to heightened arousal or even mania. Imbalances in dopamine signaling can disrupt mood regulation and contribute to mental health disorders, such as depression, anxiety, and bipolar disorder.

3. **Learning and Memory:**

    Dopamine enhances learning and memory by facilitating synaptic plasticity, the process by which neurons strengthen their connections with each other. This synaptic strengthening is necessary for the formation of long-term memories. Dopamine also helps prioritize information that is rewarding or meaningful, making it easier to remember things associated with positive experiences.

4. **Addiction and Reward-Seeking Behavior**:

   Dopamine plays a central role in the brain's addiction pathways. Drugs, alcohol, gambling, and even certain behaviors can cause a massive release of dopamine, reinforcing the desire to repeat the behavior. Over time, this can lead to addiction, as the brain becomes conditioned to seek out the reward, even at the cost of health or well-being.

Dopamine's Role in Cognitive Performance

Dopamine is crucial for many cognitive functions, including attention, working memory, and cognitive flexibility. These functions are often described as **executive functions**, which are the mental processes that enable us to plan, make decisions, solve problems, and regulate our behavior. Dopamine's involvement in these processes is why people with dopamine dysfunction (such as those with ADHD or Parkinson's disease) often experience difficulty focusing, controlling impulses, or remembering information.

Increased dopamine signaling in the prefrontal cortex, for example, can improve cognitive flexibility—our ability to adapt our thinking in response to new information. This has direct implications for problem-solving, decision-making, and learning, which are all influenced by how dopamine is utilized in the brain.

Conclusion

Dopamine is not just the brain's pleasure chemical—it's a versatile neurotransmitter involved in motivation, learning, emotional regulation, movement, and cognitive function. Understanding the science of dopamine, from its synthesis and release to its action in various brain pathways, is the first step toward mastering its effects on our mental and emotional health.

In the next chapter, we will dive deeper into the specific roles of the **DRD1** and **DRD2** receptors—how they are activated, how they influence behavior, and how their interactions shape everything from motivation to addiction. Understanding these receptors will give you the tools to harness the power of dopamine and achieve greater cognitive and emotional balance.

Chapter 3: Decoding DRD1 and DRD2 – Understanding the Genetic and Molecular Basis

Dopamine receptors are the gateways through which dopamine exerts its diverse effects on the brain and body. Among the five known types of dopamine receptors, **DRD1** and **DRD2** are the most extensively studied, due to their central role in cognition, behavior, and mental health. Understanding these receptors at both the **genetic** and **molecular** levels is essential for unlocking the power of dopamine and learning how to harness it for cognitive and emotional balance.

In this chapter, we will explore the genetic foundation of **DRD1** and **DRD2** receptors, examine how genetic variations influence receptor function, and investigate the molecular mechanisms that underlie receptor activation and inhibition. By delving into the intricate workings of these receptors, you'll gain insight into how they shape our thoughts, actions, and emotional states.

What Are DRD1 and DRD2 Receptors?

Dopamine receptors are proteins found on the surface of neurons in the brain. They belong to a family of receptors called **G-protein coupled receptors (GPCRs)**, which transmit signals inside the cell when activated by a neurotransmitter—in this case, dopamine.

DRD1 and **DRD2** are the two most significant dopamine receptors in the brain's reward system and play crucial roles in motivation, movement, learning, and emotional regulation.

- **DRD1** receptors are primarily found in the **prefrontal cortex, striatum**, and other regions associated with **cognition** and **reward processing**. They are involved in the activation of **cAMP (cyclic adenosine monophosphate)** pathways, which stimulate neurons and promote cognitive flexibility, attention, and motivation.
- **DRD2** receptors, in contrast, are predominantly located in the **striatum** and **nucleus accumbens**, areas critical for **habit formation, motor control**, and **reward reinforcement**. Activation of DRD2 receptors inhibits the cAMP pathway, leading to inhibitory signaling that regulates pleasure, motor functions, and the reinforcement of behaviors.

Both receptors are involved in brain circuits that govern how we process rewards, make decisions, and experience emotions. However, their functions differ in terms of the specific behavioral outcomes they influence.

Genetic Variations and Their Influence on Receptor Function

The genes that encode DRD1 and DRD2 receptors are highly variable, with different genetic variants leading to variations in receptor function and responsiveness. These genetic differences can influence an individual's susceptibility to mental health conditions, addiction, and cognitive performance.

DRD1 Gene Variations:

The **DRD1 gene** is located on chromosome 5 and encodes the D1 receptor. Variations in this gene can impact receptor density and signaling efficiency, which, in turn, affects brain function.

- **Gene Polymorphisms**: A common genetic variant of DRD1, known as the **-48G/A polymorphism**, has been linked to altered receptor function. This polymorphism affects the expression of DRD1 receptors, influencing **cognitive performance** and **attention**. Individuals with certain DRD1 variants may exhibit enhanced cognitive flexibility and learning abilities, while others may have difficulty regulating attention or performing complex tasks.
- **Impact on Mental Health**: Variations in DRD1 have also been associated with psychiatric conditions such as **schizophrenia** and **ADHD**. Reduced DRD1 receptor activity may contribute to cognitive deficits, poor impulse control, and attention-related problems commonly observed in these disorders.

DRD2 Gene Variations:

The **DRD2 gene**, located on chromosome 11, encodes the D2 receptor. Just like DRD1, genetic variations in the DRD2 gene can influence the receptor's functionality and impact behaviors related to addiction, reward-seeking, and motor control.

- **Gene Polymorphisms**: One of the most studied variants of the DRD2 gene is the **A1 allele** of the **Taq1A polymorphism**. People carrying this variant tend to have fewer DRD2 receptors in the brain, which has been linked to higher susceptibility to **addictive behaviors, impulsivity**, and **lower reward sensitivity**.
- **Impact on Addiction**: DRD2 receptor density plays a central role in the brain's reward system. People with lower receptor density may find it harder to feel satisfied by natural rewards (e.g., food, social interactions) and may be more prone to engaging in risky behaviors, including substance abuse. Studies have shown that the **A1 allele** is associated with increased risk for **drug addiction, alcoholism**, and **gambling disorder**.

Molecular Mechanisms of Receptor Activation and Inhibition

Understanding the molecular mechanisms that underlie the activation and inhibition of DRD1 and DRD2 receptors is key to comprehending how dopamine influences brain function and behavior.

DRD1 Receptor Activation:

When dopamine binds to DRD1 receptors, it activates a **G-protein** coupled mechanism that increases **cAMP** production. This cascade of molecular events leads to several key effects:

Mastering the balance of dopamine and its receptors is not just about curing disorders or managing negative behaviors. It's about harnessing the power of these receptors to unlock potential and foster personal growth. Whether it's enhancing your cognitive abilities, improving emotional resilience, or simply finding a better balance between work and play, understanding dopamine's role can transform the way you approach everyday life.

- **Optimizing Motivation**: By understanding how DRD1 affects motivation and goal-directed behavior, you can create strategies to enhance your productivity and success. Knowing how to activate and balance DRD1 can help you stay focused and driven, whether you're pursuing long-term goals or overcoming short-term obstacles.
- **Enhancing Learning and Memory**: Dopamine's impact on learning and memory is profound. Activating DRD1 can enhance cognitive flexibility, memory consolidation, and focus. Understanding how dopamine works can lead to better study habits, faster learning, and improved cognitive performance.
- **Regulating Emotions**: Dopamine's relationship with DRD2 can help regulate mood, decrease stress, and enhance emotional well-being. By balancing these receptors, you can master emotional regulation, achieve a healthier outlook on life, and improve your ability to deal with life's challenges.

- **Excitation of Neurons**: The increased cAMP levels activate a series of enzymes, which ultimately lead to the phosphorylation of proteins within the neuron. This process increases the excitability of the neuron, promoting cognitive flexibility, enhanced attention, and motivation.
- **Influence on Learning and Memory**: DRD1 receptor activation has been shown to enhance synaptic plasticity, the ability of synapses (the connections between neurons) to strengthen or weaken over time in response to activity. This is a fundamental mechanism of learning and memory. DRD1 receptors are particularly active in the **prefrontal cortex**, which is involved in higher cognitive functions like decision-making and working memory.
- **Role in Reward Processing**: DRD1 is also involved in the brain's reward pathways, particularly in the **mesocortical** and **mesolimbic** circuits. By modulating dopamine release in these areas, DRD1 helps drive goal-directed behavior and reward-seeking action.

DRD2 Receptor Activation:

In contrast to DRD1, activation of DRD2 receptors leads to **inhibition** of the cAMP pathway. This causes a series of events that reduce neuronal excitability and regulate behaviors related to reward, reinforcement, and motor control:

- **Inhibition of cAMP Pathway**: The activation of DRD2 receptors triggers the **Gi protein**, which inhibits the enzyme **adenylyl cyclase**, thus reducing the production of cAMP. This leads to decreased activation of protein kinase A (PKA) and reduced neuronal excitability.
- **Regulation of Reward and Reinforcement**: DRD2 plays a key role in reward processing and reinforcement learning, particularly in the **nucleus accumbens**. When dopamine binds to DRD2 receptors in this region, it dampens excessive reward-seeking behaviors and prevents the overstimulation of the reward system. In essence, DRD2 receptors help balance the brain's sensitivity to rewards, ensuring that the individual does not engage in excessive or harmful pursuit of pleasurable experiences.
- **Motor Control**: DRD2 receptors are highly concentrated in the **striatum**, an area crucial for motor function. Dopamine signaling through DRD2 regulates motor control, and its dysfunction can lead to motor impairments, as seen in **Parkinson's disease**.

The Interplay Between DRD1 and DRD2 – A Delicate Balance

While DRD1 and DRD2 receptors often function independently, they also interact in complex ways to regulate behavior. The delicate balance between the excitatory effects of DRD1 and the inhibitory effects of DRD2 ensures proper functioning of the brain's reward and motivation systems.

- **Balanced Reward Processing**: The interaction between DRD1 and DRD2 receptors in the **striatum** is essential for maintaining a healthy reward system. DRD1 receptors increase the excitability of neurons and promote goal-directed behavior, while DRD2 receptors inhibit excessive reward-seeking actions, preventing impulsivity and compulsivity.
- **Motivation and Decision-Making**: The balance between DRD1 and DRD2 activity is critical for decision-making processes. Overactivation of DRD1 can lead to **overactive pursuit of rewards**, while excessive DRD2 activation may reduce motivation or hinder goal-directed behavior. A well-balanced interaction between these two receptors ensures that we stay motivated and driven to pursue our goals without becoming overwhelmed by impulsive desires.

Conclusion

The genetic and molecular foundations of DRD1 and DRD2 receptors are integral to understanding how dopamine influences our thoughts, emotions, and behaviors. Variations in these receptors, driven by genetic factors, can shape cognitive performance, addiction susceptibility, and mental health outcomes. By understanding the molecular mechanisms of receptor activation and inhibition, we gain insight into how the brain processes rewards, motivates behavior, and regulates emotions.

In the next chapter, we will explore the specific roles of **DRD1** and **DRD2** in motivation and goal-directed behavior, providing a deeper understanding of how these receptors can be harnessed to enhance cognitive and emotional balance.

Chapter 4: The D1 Receptor – The Pursuit of Reward and Motivation

Dopamine, a key neurotransmitter in the brain's reward system, is primarily responsible for regulating motivation, pleasure, learning, and memory. The **D1 receptor (DRD1)** plays a central role in these processes, making it essential for our understanding of how motivation and goal-directed behavior are driven. In this chapter, we will explore the pivotal role of DRD1 in pursuing rewards, its influence on cognitive functions such as learning and memory, and its impact on cognitive flexibility and executive functions. By mastering the functioning of DRD1, we can better understand and optimize our drive to achieve goals, pursue personal growth, and enhance cognitive performance.

The Role of DRD1 in Motivation and Goal-Directed Behavior

At the core of motivation is the brain's ability to initiate and maintain goal-directed actions, which depend heavily on the activation of dopamine pathways. **DRD1 receptors** are especially prevalent in areas of the brain such as the **prefrontal cortex** and **striatum**, regions crucial for planning, decision-making, and reward processing.

- **Motivational Drive**: DRD1 activation in the **mesocortical** pathway stimulates motivation by facilitating the release of dopamine in response to expected rewards. This makes individuals more likely to initiate and sustain goal-directed actions.

- **Goal Pursuit**: The DRD1 receptor helps reinforce the **approach behavior** in the face of anticipated rewards. When dopamine binds to DRD1 receptors, it signals the brain to pursue positive outcomes and avoid negative ones, which are fundamental for setting and achieving long-term goals. This explains why individuals with heightened DRD1 receptor activity tend to show greater persistence and resilience when pursuing difficult tasks or long-term objectives.

- **Dopaminergic Influence on Behavior**: DRD1 is instrumental in shaping **reward-seeking behavior**. This receptor's activation leads to increased neuronal firing in areas like the **striatum** and **nucleus accumbens**, facilitating the selection of goal-directed behaviors that maximize positive rewards. Therefore, individuals with a higher density of DRD1 receptors are more likely to engage in behaviors that lead to tangible rewards or success.

How DRD1 Activation Influences Learning and Memory

A significant consequence of DRD1 receptor activation is its influence on **learning** and **memory** processes, particularly in how the brain responds to new information and integrates experiences to improve decision-making. DRD1 enhances **synaptic plasticity**, which is the brain's ability to form and strengthen connections between neurons based on experience.

- **Memory Consolidation**: DRD1 receptors help solidify information into long-term memory, a process known as **memory consolidation**. When dopamine is released and binds to DRD1, it facilitates the stabilization of new memories. This is why dopamine levels, and by extension DRD1 activity, are closely linked to the strength and durability of memory formation.

- **Learning Enhancement**: DRD1 receptors promote learning by modulating **synaptic plasticity** in the brain's **hippocampus** and **prefrontal cortex**. These regions are involved in encoding new information and adapting behavior based on past experiences. DRD1 activation enhances the brain's ability to learn from rewards and outcomes, reinforcing the actions that led to successful results. Thus, a higher DRD1 receptor density is often associated with superior learning abilities and more efficient memory retention.

- **Adaptive Learning**: DRD1 is also involved in **adaptive learning**, which allows the brain to adjust behavior based on changing environmental conditions. It plays a role in **cognitive flexibility**—the ability to switch between tasks, strategies, and thought patterns. For instance, DRD1 helps individuals modify their approach if their current strategy is not yielding the desired reward.

Impact of DRD1 on Cognitive Flexibility and Executive Functions

Cognitive flexibility refers to the brain's capacity to adapt to new situations and adjust thinking or behavior in response to changing goals. This is a critical component of **executive functions**, which include high-level cognitive abilities like **decision-making**, **problem-solving**, and **planning**. DRD1 plays a vital role in regulating these functions, and its activation helps optimize mental performance in dynamic and uncertain environments.

- **Cognitive Flexibility**: DRD1 is implicated in the brain's ability to shift from one task to another, learn from past mistakes, and explore alternative strategies. This adaptability is essential in real-world scenarios where circumstances change rapidly and require flexible thinking to achieve desired outcomes.

- **Executive Control**: In the prefrontal cortex, DRD1 receptors help regulate **working memory**, which is the ability to temporarily hold and manipulate information. Working memory is essential for effective decision-making, problem-solving, and the capacity to plan complex tasks. Higher activation of DRD1 enhances one's ability to retain and manipulate relevant information, increasing the brain's efficiency in high-level cognitive tasks.

- **Decision-Making and Risk Assessment**: Cognitive flexibility, in turn, allows individuals to make decisions with a better understanding of the potential outcomes and consequences of their actions. DRD1 is involved in the **evaluation of rewards** and **risk assessment**. When faced with a choice, DRD1 activity helps the brain weigh the potential benefits and risks, promoting more effective decision-making.

- **Executive Function Disorders**: Reduced DRD1 function or receptor density can lead to deficits in executive functions, including difficulties with planning, self-control, and switching between tasks. These deficits are often seen in individuals with **attention-deficit hyperactivity disorder (ADHD), schizophrenia**, and other cognitive disorders.

Enhancing DRD1 Activity for Cognitive and Emotional Benefits

Given its influence on motivation, learning, cognitive flexibility, and executive functions, optimizing DRD1 activity has broad implications for personal growth, academic success, and emotional regulation. Here are some ways to enhance DRD1 activity:

- **Physical Exercise**: Regular physical activity, especially aerobic exercise, has been shown to increase the availability of dopamine in the brain, including enhancing the function of DRD1 receptors. Exercise improves brain plasticity and cognitive function, promoting neurogenesis (the growth of new neurons) and enhancing memory and learning capacity.
- **Mental Stimulation**: Engaging in cognitively challenging activities, such as puzzles, learning a new language, or playing a musical instrument, can stimulate DRD1 receptor activity. These activities promote neuroplasticity and the development of new neural connections, reinforcing the brain's ability to learn and adapt.
- **Mindfulness and Meditation**: Practices like mindfulness meditation have been shown to modulate dopamine levels and can enhance the function of DRD1 receptors. These practices help improve emotional regulation, increase cognitive flexibility, and reduce stress, all of which support a balanced dopamine system.
- **Nutrition and Supplements**: Certain nutrients and supplements, such as **tyrosine** (the precursor to dopamine), **omega-3 fatty acids**, and **L-theanine**, may help optimize dopamine signaling, including DRD1 receptor function. A healthy, balanced diet rich in these nutrients supports the brain's ability to produce and utilize dopamine efficiently.

Conclusion

The **DRD1 receptor** is a critical player in the pursuit of rewards, motivation, and goal-directed behavior. It shapes our cognitive and emotional experiences by influencing learning, memory, decision-making, and cognitive flexibility. By enhancing DRD1 receptor function, we can improve our ability to stay focused on goals, adapt to new situations, and learn more effectively. This not only helps with personal growth and achievement but also supports emotional regulation and mental clarity.

As we continue to explore the **D2 receptor** in the next chapter, we will see how it works in tandem with DRD1 to regulate pleasure, habit formation, and reinforcement, providing a holistic understanding of the brain's complex reward systems.

Chapter 5: The D2 Receptor – Regulation of Pleasure and Habit Formation

Dopamine is at the heart of the brain's reward and reinforcement systems, and the **D2 receptor (DRD2)** plays a central role in modulating pleasure, motivation, and habit formation. While DRD1 is often associated with goal-directed behavior, DRD2 is more involved with **pleasure** and **reinforcement**, making it critical for understanding the balance between reward-seeking and the formation of habits, including those related to addiction. In this chapter, we will explore the functions of the D2 receptor, its relationship with **reinforcement learning**, and its role in **addiction** and **motor control**, particularly in conditions like **Parkinson's disease**.

DRD2's Role in Reinforcement and Addiction

The DRD2 receptor plays an essential role in the brain's reinforcement learning system, where it regulates how we experience pleasure and form habits based on reward outcomes. It is predominantly found in areas of the brain that are involved in **reward processing**, including the **nucleus accumbens**, **ventral tegmental area**, and **striatum**.

- **Pleasure and Reward**: When a rewarding stimulus—whether it's food, social interaction, or a substance—activates dopamine release, DRD2 receptors help signal pleasure and satisfaction. The greater the activation of DRD2, the more intense the pleasure, reinforcing the behavior that led to the reward. This reinforcement loop strengthens the connection between a specific action and its associated reward.
- **Habits and Compulsions**: Over time, the constant reinforcement of pleasurable actions leads to habit formation. DRD2 receptors are instrumental in the process of **habitual behavior**: when an activity is repeatedly associated with positive reinforcement, it becomes ingrained, and the brain starts to anticipate the reward. As these habits become more entrenched, they can evolve into compulsions—behaviors that are difficult to control even in the absence of immediate rewards.

- **Addiction and Dopamine Dysregulation**: In the context of addiction, DRD2 is crucial for understanding how substances or behaviors hijack the brain's reward system. Addictive drugs (like cocaine, alcohol, or opioids) artificially elevate dopamine levels, leading to overstimulation of DRD2 receptors. This overstimulation can cause **dopamine dysregulation**, impairing the brain's natural reward mechanisms. Over time, the brain becomes dependent on these external stimuli to trigger the same level of pleasure, leading to **tolerance** (where the same reward no longer produces the same effect) and **addiction** (compulsive engagement in the behavior despite negative consequences).

The Relationship Between DRD2 and Pleasure-Seeking Behavior

While DRD1 regulates the drive to pursue goals, **DRD2 governs the experience of pleasure itself**. The balance between these two receptors is crucial for understanding the way we approach rewards and how we derive satisfaction from them. Too little or too much activation of either receptor can lead to a variety of mental health issues, including **anxiety**, **depression**, or **compulsive behavior**.

- **Pleasure-Seeking vs. Self-Control**: DRD2 is also involved in moderating the level of pleasure we derive from different activities. Those with **lower DRD2 receptor density** may experience diminished pleasure from natural rewards (like socializing, exercising, or eating), which can lead them to seek out more intense or artificial sources of stimulation, such as drugs, gambling, or other compulsive behaviors. On the other hand, individuals with **higher DRD2 density** tend to have a more balanced relationship with pleasure and may be better able to regulate their reward-seeking behaviors.

- **Pleasure and Risk**: The relationship between DRD2 and pleasure-seeking behavior also extends to risk-taking. **High DRD2 activity** has been linked to **novelty-seeking** and the willingness to take risks for the promise of a reward. This is particularly relevant in contexts such as entrepreneurship, gaming, and social interactions, where the anticipation of a reward (e.g., money, success, or attention) can drive people to engage in high-risk activities. Conversely, low DRD2 activity may make individuals more risk-averse or less motivated to seek rewards, which could lead to decreased pleasure and motivation.

The Role of DRD2 in Motor Control and Parkinson's Disease

In addition to its role in pleasure, DRD2 is also crucial for **motor control**. Its dysfunction is closely associated with **Parkinson's disease**, a neurodegenerative disorder that affects movement.

- **Motor Control and Movement Disorders**: DRD2 receptors are abundant in the **striatum**, an area of the brain involved in coordinating voluntary movement. Dopamine released into the striatum activates DRD2, which helps facilitate smooth and controlled movement. In **Parkinson's disease**, the progressive degeneration of dopamine-producing neurons leads to **reduced dopamine signaling**, particularly at the D2 receptor level. This results in the motor symptoms characteristic of Parkinson's, such as tremors, rigidity, bradykinesia (slowness of movement), and postural instability.

- **Therapeutic Targeting of DRD2**: Medications that target the DRD2 receptor, such as **dopamine agonists** and **L-DOPA** (a precursor to dopamine), are commonly used to manage the symptoms of Parkinson's disease. These treatments aim to restore dopamine function and stimulate DRD2 receptors, improving motor control and reducing symptoms of the disease. However, **long-term use** of these medications can sometimes lead to side effects like **dopamine dysregulation syndrome**, which involves compulsive behaviors such as gambling or excessive shopping.

DRD2's Influence on Reward and Motivation Pathways

The interplay between DRD1 and DRD2 creates a nuanced system for regulating motivation, reward, and habit formation. DRD1 primarily enhances motivation for goal-directed behavior, while DRD2 plays a key role in modulating the actual pleasure or satisfaction derived from rewards. This dynamic is essential for understanding how the brain evaluates potential rewards and how it drives behavior in response to those rewards.

- **Neuroplasticity and Habit Formation**: DRD2's influence on reinforcement learning means that the receptor plays a key role in the process of **neuroplasticity**—the brain's ability to adapt based on experience. By strengthening the neural pathways associated with certain behaviors, DRD2 facilitates the formation of habits. While these habits can be positive (e.g., regular exercise, productive work habits), they can also become negative, especially when driven by addictive behaviors.

- **Emotional Regulation and Reward Sensitivity**: DRD2 is also involved in regulating emotional responses to rewards. It affects how the brain responds to emotional stimuli and plays a role in the balance between **pleasure and discomfort**. This makes DRD2 an essential receptor for emotional regulation and resilience. Understanding how to optimize DRD2 function can help individuals maintain a balanced emotional state, particularly when navigating the complexities of stress and reward-seeking behavior.

Enhancing DRD2 Activity for Emotional and Behavioral Benefits

Given its role in addiction, pleasure-seeking behavior, and motor control, optimizing DRD2 function can have significant benefits for emotional well-being and habit formation. Here are some ways to enhance DRD2 activity:

- **Exercise and Movement**: Physical activity, especially aerobic exercises like running or cycling, has been shown to increase dopamine release and improve DRD2 receptor sensitivity. Exercise enhances **dopamine signaling** in the brain, which not only improves motor function but also contributes to a sense of well-being and emotional balance.
- **Social Interactions**: Positive social interactions also stimulate dopamine release and activate DRD2 receptors. Engaging in meaningful relationships, having a strong social support system, and participating in enjoyable social activities can help regulate reward-seeking behavior and enhance emotional satisfaction.
- **Nutrition**: Certain foods, such as those rich in **omega-3 fatty acids** (found in fish like salmon and in walnuts) and **tyrosine** (found in lean meats, dairy, and soy), can help support healthy dopamine function and enhance DRD2 activity. These nutrients support the production of dopamine and the functioning of dopamine receptors, promoting a healthy balance between reward and motivation.
- **Mindfulness and Meditation**: Research suggests that mindfulness practices, such as **meditation** and **deep breathing exercises**, can help regulate dopamine levels and enhance emotional well-being. By promoting a state of calm and focus, these practices help balance the activation of dopamine receptors, including DRD2, leading to improved emotional regulation and reduced compulsive behaviors.

Conclusion

The **DRD2 receptor** plays a critical role in regulating pleasure, motivation, and habit formation. By understanding the mechanisms of DRD2, we can gain insights into how the brain evaluates rewards and forms both adaptive and maladaptive behaviors. Whether it's reinforcing healthy habits, overcoming addictive behaviors, or managing emotional responses to rewards, optimizing DRD2 function is essential for achieving a balanced and fulfilling life.

As we move into the next chapter, we will explore how the dynamic interaction between DRD1 and DRD2 receptors shapes decision-making, impulsivity, and emotional regulation, offering further insights into how to master dopamine balance for optimal mental health and well-being.

Chapter 6: The Dynamic Interaction Between DRD1 and DRD2

The human brain is a highly intricate network of circuits that govern our behavior, emotions, and cognitive processes. At the heart of many of these circuits are two key players in the dopamine system: **DRD1** and **DRD2** receptors. While these receptors function independently in some contexts, their **dynamic interaction** is crucial for the balance of motivation, pleasure, reward, and learning. This chapter explores how **DRD1** and **DRD2** receptors work together to regulate brain function, their role in mental health, and how an imbalance between the two can affect decision-making, impulsivity, and emotional regulation.

The Interplay Between DRD1 and DRD2 in the Brain's Reward System

Both **DRD1** and **DRD2** receptors are found in areas of the brain involved in **reward processing, motivation**, and **emotion regulation**. These receptors are part of the brain's **dopaminergic pathways**, primarily the **mesolimbic, mesocortical**, and **nigrostriatal** systems. Their combined function is essential for the brain to evaluate and respond to rewards, goals, and environmental cues.

Complementary Roles in Motivation and Reward

- **DRD1** is typically associated with the **initiation of goal-directed behavior**, especially in the pursuit of **novel rewards**. It is crucial for motivation, cognitive flexibility, and the ability to adapt to changing circumstances.
- **DRD2**, on the other hand, modulates the brain's response to **pleasure** and **reinforcement**. It is more concerned with the **evaluation** of reward and reinforcement learning, determining how rewarding a behavior or outcome feels and how likely it is to be repeated in the future.

2. Together, these receptors form a feedback loop that influences the **initiation**, **pursuit**, and **reinforcement** of goal-directed actions. In this way, DRD1's role in motivation aligns with DRD2's role in reward processing to facilitate effective behavior. The balance between the two determines whether the brain will find an experience rewarding enough to repeat or whether motivation will wane over time.

3. **Cross-talk Between DRD1 and DRD2** While DRD1 and DRD2 receptors are typically thought of in isolation, they do not operate independently. These two receptors often exhibit **functional cross-talk**, meaning they can influence each other's activity. For instance, the activation of one receptor can affect the **sensitivity** and **activity** of the other, resulting in either **amplified or diminished effects** on behavior and cognition.

- **DRD1 activation** can modulate DRD2 activity in areas such as the **nucleus accumbens**, an area central to motivation and reward. Conversely, DRD2 activity can influence the signaling of DRD1, thus affecting the **motivation to engage in rewarding behaviors**.
- This interaction between DRD1 and DRD2 is particularly important for the brain's ability to navigate **ambiguous** or **complex** reward scenarios. For example, in situations requiring **decision-making under uncertainty**, the dynamic balance between DRD1 and DRD2 helps the brain weigh the potential rewards of different options and choose the most beneficial course of action.

The Impact of DRD1 and DRD2 Imbalance on Mental Health

An imbalance between DRD1 and DRD2 activity can have significant implications for mental health. In certain psychological and neurological disorders, this dysregulation can lead to problems with **impulsivity, emotional regulation**, and **decision-making**. Understanding how the two receptors influence each other provides insight into these disorders and potential avenues for intervention.

Imbalance and Impulsivity

- An imbalance favoring **DRD1 activation** might lead to excessive motivation without a corresponding level of reward processing, resulting in heightened impulsivity and **poor decision-making**. For instance, people with higher DRD1 activity may pursue goals aggressively, but if their **DRD2 function** does not match that level of motivation, they might not feel sufficiently rewarded, which can lead to frustration, risk-taking behavior, and even addiction.
- Conversely, **too much DRD2 activity** can lead to excessive reward-seeking, making the individual overly sensitive to **rewards and reinforcements**. This can result in a **compulsive desire** to repeat certain behaviors, even if they are not conducive to long-term well-being.

Decision-Making and Cognitive Flexibility

- The **balance** between DRD1 and DRD2 influences cognitive flexibility—the ability to adapt to changing circumstances and switch strategies when needed. A **dominance of DRD1** activity can make it difficult for an individual to adjust to new information, as they become fixated on their current course of action. Meanwhile, a **dominance of DRD2** activity can lead to being overly cautious and reluctant to take risks, which can impede **innovative thinking** and **problem-solving**.
- For instance, individuals with **Parkinson's disease**, where there is often a **reduction in DRD2 activity**, may struggle with cognitive flexibility, often becoming rigid in their thinking or behaviors. Alternatively, those with certain forms of **bipolar disorder** may exhibit extreme fluctuations in reward sensitivity, where **high DRD1 activity** during manic episodes drives excessive goal-seeking, while **low DRD2 activity** leads to a reduced sense of pleasure during periods of depression.

Emotional Regulation

- DRD1 and DRD2 imbalances can also impact how individuals regulate their emotions. **Low DRD2 activity**, for example, is associated with **depression** and **anhedonia** (the inability to feel pleasure), where the brain does not appropriately process rewards or positive emotions. On the other hand, **high DRD1 activity** may lead to **overactivity in the pursuit of goals**, causing the individual to become overly driven and emotionally reactive when goals are not achieved.
- In contrast, an optimal balance between DRD1 and DRD2 would enable individuals to **pursue goals effectively** while also maintaining **emotional equilibrium**, experiencing rewards in a healthy, regulated manner.

Implications for Impulsivity and Self-Control

The interplay between DRD1 and DRD2 has profound implications for how we manage **impulsivity** and **self-control**. **High levels of impulsivity** are often associated with an overactive DRD1 system and an underactive DRD2 system. For instance, individuals who struggle with **addiction** or **compulsive behaviors** may experience a **high drive to seek rewards (DRD1)** but a **low reward response (DRD2)**, leading them to continually seek new stimuli without feeling satiated.

The dynamic between DRD1 and DRD2 also informs our ability to exercise **self-control**. DRD1 influences **motivation**, while DRD2 influences **reward sensitivity**. A delicate balance allows for **appropriate goal-setting** and the ability to **delay gratification**—an essential skill for overcoming impulsive urges and developing healthy long-term habits.

To optimize this dynamic:

- **Mindfulness techniques** can help regulate the activation of these receptors by encouraging present-moment awareness, which balances both **impulsive drives** (DRD1) and **reward sensitivity** (DRD2).
- **Cognitive-behavioral therapy (CBT)**, which focuses on altering patterns of thought and behavior, can help individuals manage the impulses driven by DRD1 and enhance their **reward regulation** driven by DRD2.

The Future of DRD1 and DRD2 Modulation

Understanding the balance between DRD1 and DRD2 opens exciting possibilities for future therapies aimed at enhancing **mental health, cognitive performance**, and **behavioral outcomes**. Researchers are increasingly focused on **targeted therapies** that can modulate the **activity** of these receptors, either through **medication, neurofeedback**, or even **genetic interventions**.

For example, individuals who have difficulty with **motivation and goal-directed behavior** may benefit from **DRD1-boosting therapies** that enhance cognitive flexibility and executive function. On the other hand, those struggling with **compulsive behaviors** may benefit from therapies that increase **DRD2 activity**, improving their ability to **experience reward** and regulate their emotional responses.

In the next chapter, we will delve deeper into how **dopamine signaling** influences **neuroplasticity**, and how the mastery of DRD1 and DRD2 can be used to create long-term changes in behavior, habits, and mental health.

Conclusion

The interaction between DRD1 and DRD2 receptors is a cornerstone of how our brain processes motivation, pleasure, and reward. These two receptors, though distinct in their functions, work together to create a dynamic system that shapes our behavior, emotions, and cognitive abilities. By understanding how these receptors interact, we gain insights into how to regulate **impulsivity, emotional balance**, and **goal-directed action**. Striking the right balance between DRD1 and DRD2 activity is essential for achieving **mental clarity, emotional regulation**, and **overall well-being**.

Chapter 7: Dopamine and the Brain's Reward System – Neuroplasticity and Behavior Change

The ability of the brain to adapt and change is one of its most remarkable features. **Neuroplasticity**, the brain's capacity to reorganize itself by forming new neural connections, is the foundation for **learning, memory**, and **behavior change**. Dopamine, through its interaction with **DRD1** and **DRD2** receptors, plays a pivotal role in **neuroplasticity** and is integral to how we form new habits, adapt to new environments, and break old patterns.

This chapter explores how **dopamine signaling**, via the modulation of **DRD1** and **DRD2** receptors, drives **long-term behavioral changes**. It also highlights how we can harness this process to **optimize habits, break addictions**, and **reinforce positive behavior**. By mastering dopamine receptors, we can unlock the power of neuroplasticity and take charge of our personal transformation.

Neuroplasticity: The Brain's Capacity for Change

Neuroplasticity refers to the brain's ability to form new **neural connections** and reorganize existing ones in response to learning, experience, or injury. This process underpins our ability to **adapt to new environments, learn new skills**, and even recover from brain injuries.

Neuroplasticity occurs in several key ways:

1. **Synaptic plasticity** – Strengthening or weakening of connections between neurons based on activity and experience.
2. **Structural plasticity** – Changes in the physical structure of the brain, such as the growth of new neurons (neurogenesis) or the formation of new synaptic connections.

Dopamine is one of the central players in neuroplasticity, especially when it comes to **reward-driven learning**. Both **DRD1** and **DRD2** receptors are involved in signaling that promotes synaptic and structural changes in the brain. When you engage in a behavior that is rewarding, dopamine is released, and it activates these receptors, which in turn facilitates **learning** and **memory consolidation**.

DRD1 and Neuroplasticity

- **DRD1** plays a crucial role in learning **new information** and adapting to changing environments. It facilitates **cognitive flexibility**, the ability to switch between different tasks or strategies. When DRD1 is activated, it enhances **synaptic plasticity** in brain regions like the **prefrontal cortex** and **hippocampus**, which are involved in decision-making, working memory, and learning.
- DRD1 activation also supports the formation of **long-term memories** related to goal-directed behavior, reinforcing the association between reward and effort. This makes **goal-setting** and **behavioral persistence** more effective.

DRD2 and Neuroplasticity

- DRD2 is involved in the brain's **reinforcement** pathways. When activated, it enhances the brain's ability to process and store experiences that are **rewarding** or **pleasurable**, thereby increasing the likelihood of repeating those behaviors in the future.
- DRD2 activation strengthens the **reward circuit**, facilitating the formation of **habitual behaviors** and the reinforcement of behavior patterns. This is particularly important in the development of both **healthy habits** and **addictive behaviors**.

By understanding the interaction between these receptors, we can leverage **dopamine signaling** to create lasting behavioral changes.

How DRD1 and DRD2 Influence Long-Term Behavior Changes

The dynamic relationship between **DRD1** and **DRD2** receptors shapes how we learn, adapt, and reinforce behavior over time. Their interaction directly influences how we develop new **habits** and how we can change ingrained patterns of behavior.

Habit Formation and Reinforcement

- **DRD1** and **DRD2** are integral to the process of habit formation. **Habits** are formed when behaviors that are consistently rewarded become automatic. DRD1 helps us initiate and pursue new behaviors, while DRD2 ensures that we feel rewarded for those behaviors. Over time, this creates a cycle of reinforcement where the brain reinforces the neural connections related to the behavior.
- For example, when we commit to a new exercise routine, **DRD1** activates to help us initiate the behavior and stay motivated, while **DRD2** reinforces the pleasure we feel from the positive outcomes (such as improved health or mood). Over time, this combination makes the behavior habitual, and it becomes an automatic part of our routine.

Breaking Addictions

- Just as dopamine signaling helps form new habits, it also plays a central role in the formation and persistence of **addictions**. In addiction, the **reward pathways** are hijacked, and the brain becomes overly sensitized to the stimuli that trigger the addictive behavior. The balance between **DRD1** and **DRD2** can influence how easily the brain forms new associations, including those associated with addiction.
- **DRD1** may drive the initial craving or desire to engage in addictive behaviors, while **DRD2** reinforces the reward associated with these behaviors. Over time, however, **neuroplastic changes** can occur that make it difficult to break these habits.
- To break addiction, we need to **reprogram** the brain's reward system. This can be done by resetting dopamine pathways and strengthening healthier, more positive behaviors. **Mindfulness, cognitive-behavioral therapy (CBT)**, and other therapeutic interventions can help balance dopamine levels and reverse the dysregulation associated with addiction.

Behavioral Flexibility and Change

- **DRD1** plays a key role in **cognitive flexibility**, which is the ability to shift strategies and adapt to new information. This is crucial for behavior change, as it enables individuals to break free from old patterns of behavior and make more adaptive choices.
- **DRD2**, on the other hand, provides the **reward** that reinforces the decision to change behavior. Together, the receptors create a balance between **motivation** (DRD1) and **reward reinforcement** (DRD2), which helps an individual maintain long-term behavior change.

Practical Applications in Habit Formation and Breaking Addictions

Understanding the roles of **DRD1** and **DRD2** in neuroplasticity and behavior change can help us **optimize habits, break addictions**, and **reinforce positive behavior**. Here are practical strategies to utilize this knowledge:

Optimize Habit Formation

- Focus on **rewarding** yourself for small, incremental steps toward your goals. This triggers **DRD2** and reinforces the behavior. For example, if you want to develop a regular exercise routine, reward yourself after each session with something pleasurable, like relaxation or a healthy treat.
- Engage in **goal-setting** that gradually increases in difficulty. **DRD1** will help you stay motivated, while **DRD2** will reinforce your progress as you achieve your milestones.
- **Consistency** is key. The more frequently a behavior is rewarded, the stronger the neural connections will become. Repetition helps reinforce the pathway that involves **DRD1** and **DRD2**, making the behavior automatic over time.

Break Addictions

- **Cognitive-behavioral therapy (CBT)** and other therapeutic techniques can help **reset the reward pathways**. By focusing on reframing thoughts and behaviors, you can alter the balance between **DRD1** and **DRD2**, breaking the addiction cycle.
- **Gradual exposure** to healthy alternatives and rewards can help reprogram the brain. For example, if someone is trying to quit smoking, introducing a healthier, rewarding behavior (such as exercise) can activate **DRD1** and **DRD2** in a positive context, reinforcing new, healthier habits.
- **Mindfulness practices** can help **calm the brain's craving for reward** and improve emotional regulation. By reducing the overactivity of **DRD1** and **DRD2** during moments of temptation, mindfulness can help rewire the brain to respond more appropriately to rewards.

Enhance Cognitive Flexibility

- Challenge yourself to regularly engage in activities that require **mental adaptation**—such as learning a new skill, switching between tasks, or solving complex problems. These activities engage **DRD1** and promote cognitive flexibility, helping you adapt to new situations and make better decisions.
- **Exposure to novelty** also stimulates **DRD1**, encouraging you to explore new possibilities and make new connections.

Conclusion

The connection between dopamine signaling and **neuroplasticity** provides the key to understanding how behavior change occurs. By mastering the interplay between **DRD1** and **DRD2**, we can enhance our ability to form healthy habits, break free from negative patterns, and reinforce behaviors that promote well-being. The brain's capacity for change is immense, and by leveraging dopamine signaling, we can unlock its full potential to achieve personal transformation and sustainable behavior change. The next chapter will dive deeper into the impact of **DRD1** and **DRD2** on mental health, specifically exploring how dysfunction in these receptors is linked to common neuropsychological disorders.

Chapter 8: DRD1, DRD2, and Mental Health – A Deep Dive into Neuropsychological Disorders

Dopamine is often referred to as the brain's "feel-good" neurotransmitter, playing a crucial role in motivation, pleasure, and reward. However, when dopamine signaling is disrupted, it can contribute to a wide range of **neuropsychological disorders**. The delicate balance between **DRD1** and **DRD2** receptors is key to understanding how dopamine influences mood, cognition, and behavior. Dysfunction in these receptors has been implicated in several major mental health conditions, including **schizophrenia**, **depression**, **bipolar disorder**, and **anxiety** disorders. This chapter will explore the intricate relationship between dopamine receptors and mental health, shedding light on how **DRD1** and **DRD2** imbalances can lead to psychiatric disorders and offering insights into potential therapeutic interventions.

The Connection Between Dopamine Dysfunction and Psychiatric Disorders

Both **DRD1** and **DRD2** play critical roles in regulating **mood, emotional responses, cognition**, and **motivation**. Alterations in the functioning of these receptors have been linked to several **psychiatric disorders**. While **DRD1** influences cognitive flexibility, learning, and reward processing, **DRD2** plays a central role in emotional regulation, habit formation, and the reinforcement of behavior.

Schizophrenia

psychotic disorder

DRD2

- **Overactivity of DRD2** in certain brain regions, especially the **mesolimbic pathway**, is thought to contribute to the positive symptoms of schizophrenia, such as **hallucinations** and **delusions**.
- Conversely, **underactivity of DRD1** in the **prefrontal cortex** has been linked to **cognitive deficits** and **negative symptoms**, such as **apathy, anhedonia**, and **difficulty with executive functioning**.
- Antipsychotic medications that block **DRD2 receptors** are commonly used to treat schizophrenia, although they can also cause side effects, such as **motor dysfunction** and **tardive dyskinesia**.

Depression

sadness

loss of interest

cognitive impairments

DRD1

DRD2

- **Hypodopaminergic activity**, or reduced dopamine signaling, has been associated with **low motivation, anhedonia** (inability to experience pleasure), and **cognitive sluggishness**, all hallmark symptoms of depression.
- **DRD1** dysfunction may impair cognitive flexibility and the ability to initiate goal-directed behavior, contributing to the **psychomotor slowing** seen in depressive states.
- Conversely, **overactivation of DRD2** in certain pathways may contribute to mood instability and **emotional dysregulation**, which are often observed in depression.

Bipolar Disorder

extreme mood swings

mania

hypomania

depression

DRD1

DRD2

- During manic episodes, **excessive dopamine release** and **overactivation of DRD1** in the **prefrontal cortex** can lead to **impulsivity, grandiosity**, and **reward-seeking behavior**.
- Conversely, during depressive episodes, **underactivity of DRD2** in areas like the **striatum** may contribute to **low energy, fatigue**, and **reduced pleasure in activities**.
- The **dopaminergic imbalance** between **DRD1** and **DRD2** is thought to contribute to the **cycle of mood shifts**, with dopamine's role in **reward processing** and **emotional regulation** being disrupted.

Anxiety Disorders

generalized anxiety disorder (GAD)

social anxiety

panic disorder

emotion regulation

- **Overactivation of DRD1** in stress-related pathways may exacerbate anxiety symptoms by increasing **hyperactivity** and **rumination**.
- On the other hand, **underactivity of DRD2** in the **prefrontal cortex** has been linked to **difficulty in emotional regulation**, making it harder to manage feelings of anxiety and stress.
- In some cases, a **dopamine deficiency** in areas of the brain responsible for **fear extinction**, such as the **amygdala**, may impair the ability to regulate fearful responses, thus contributing to **chronic anxiety**.

The Role of Dopamine in Anxiety, Stress, and Emotional Regulation

Dopamine is intricately involved in the regulation of **emotional responses** and the brain's reaction to **stressful situations**. Both **DRD1** and **DRD2** receptors influence how we process emotional stimuli, regulate our reactions, and experience stress.

Stress Response

stress response

HPA axis

- **DRD1** is involved in regulating the **prefrontal cortex**, an area critical for **decision-making, cognitive flexibility**, and **self-regulation**. Dysfunction in this receptor can impair the ability to manage stress effectively, leading to **increased emotional reactivity**.
- **DRD2** affects **reward processing** and **pleasure regulation**, which is particularly important in how we cope with stress. Dysregulation of **DRD2** signaling may contribute to **depressive symptoms**, as individuals may have difficulty experiencing pleasure in normal activities or rewards.

Emotion Regulation

balance between DRD1 and DRD2

- **DRD1 dysfunction** can result in **cognitive rigidity**, making it difficult to adapt to changing emotional states and leading to difficulties in problem-solving during emotionally charged situations.
- **DRD2 dysfunction**, on the other hand, is linked to **difficulty in experiencing reward** and **pleasure** in emotionally neutral or positive events, contributing to **emotional blunting** and **depression**.

Therapeutic Implications – Targeting Dopamine Dysfunction in Mental Health

Understanding the role of **DRD1** and **DRD2** in mental health opens up new avenues for therapeutic interventions. Pharmacological and non-pharmacological strategies that target dopamine receptors can help balance the reward system, improve **mood regulation**, and alleviate symptoms of **neuropsychological disorders**.

Pharmacological Interventions

- **Antipsychotic Medications**: These are commonly used in the treatment of disorders like schizophrenia and bipolar disorder. By blocking **DRD2** receptors, these drugs help reduce the **overactivation of dopamine** in certain brain regions, alleviating **hallucinations, delusions**, and other positive symptoms. However, **side effects** such as **movement disorders** may occur, especially with long-term use.
- **Dopamine Agonists and Antagonists**: Drugs that selectively activate or block **DRD1** and **DRD2** can be used to adjust dopamine levels in the brain. For instance, **dopamine agonists** may be useful for **depression** or **cognitive dysfunction** associated with conditions like Parkinson's disease, while **dopamine antagonists** can reduce the symptoms of mania or psychosis.
- **Selective Serotonin and Norepinephrine Reuptake Inhibitors (SSRIs/SNRIs)**: While these drugs primarily target serotonin and norepinephrine, they can also have indirect effects on dopamine levels, contributing to mood stabilization and reduced anxiety.

Non-Pharmacological Approaches

- **Cognitive Behavioral Therapy (CBT)**: CBT helps individuals manage negative thought patterns and reframe maladaptive behaviors. By **improving emotional regulation** and promoting healthier coping strategies, CBT can help recalibrate the dopamine system and restore balance between **DRD1** and **DRD2**.
- **Mindfulness and Meditation**: These practices help regulate dopamine release, reduce **emotional reactivity**, and improve **stress management**, thereby contributing to better mental health and emotional stability.

Conclusion

Dopamine, through its action on **DRD1** and **DRD2**, plays a foundational role in regulating emotions, cognition, and behavior. When these receptors are dysregulated, mental health conditions like **schizophrenia, depression, bipolar disorder**, and **anxiety** can arise. By understanding the molecular mechanisms behind these disorders, we can better treat and manage symptoms through both **pharmacological and non-pharmacological interventions**. Optimizing dopamine function offers promising pathways for improving mental health and enhancing cognitive and emotional well-being, ultimately leading to a more balanced and fulfilling life.

Chapter 9: Dopamine Dysregulation – The Science of Addiction

Addiction is one of the most pervasive and complex challenges facing modern society, with profound impacts on individuals, families, and communities. At the core of addiction lies the brain's **dopamine system**, which plays a central role in the experience of **reward, motivation**, and **pleasure**. When this system is dysregulated, it can lead to compulsive behavior, loss of control, and the development of **addiction**. This chapter explores how addiction hijacks the dopamine system, with a focus on the roles of the **DRD1** and **DRD2** receptors. It also delves into how individuals can restore balance to their dopamine systems to break free from the cycle of addiction and reclaim control over their lives.

The Dopamine System and Addiction

The brain's **dopamine system** is intricately involved in reinforcing behavior that is perceived as rewarding. When we engage in activities that are pleasurable or beneficial for survival (such as eating, socializing, or reproducing), dopamine is released as part of the **brain's reward pathway**, creating feelings of pleasure and satisfaction. This **dopamine release** encourages repetition of the behavior, ensuring survival and promoting well-being.

However, the dopamine system can be **hijacked** by addictive substances or behaviors. Drugs like **cocaine**, **amphetamines**, **nicotine**, and **alcohol**, as well as addictive behaviors such as **gambling** or **social media use**, can cause an unnatural surge in dopamine levels, leading to exaggerated feelings of pleasure or euphoria. Over time, this constant overstimulation of the dopamine system can alter the functioning of the brain's reward pathways, leading to **neuroadaptation**—the brain's attempt to adjust to the repeated overstimulation.

In the long run, this dysregulation can cause the **dopamine system** to become less responsive to natural rewards. This leads to a cycle where the individual becomes dependent on the substance or behavior to trigger dopamine release, while simultaneously losing the ability to experience pleasure from other, more natural sources. This is the essence of **addiction**: compulsively seeking out the reward (whether substance or behavior) to avoid feelings of discomfort or lack of motivation.

The Roles of DRD1 and DRD2 in Addiction

Both **DRD1** and **DRD2** receptors are critical players in addiction, but they operate in slightly different ways. Their roles are crucial not only in the development of addiction but also in its persistence and the challenges faced during recovery.

The Role of DRD1 in Addiction

- **DRD1** receptors are primarily involved in **motivation, goal-directed behavior**, and the **pursuit of rewards**. When DRD1 receptors are activated, they enhance the motivation to seek out rewards, including addictive substances or behaviors.
- **Overactivation of DRD1** in certain regions of the brain, such as the **prefrontal cortex** and **striatum**, can lead to heightened impulsivity and an exaggerated response to rewards, making an individual more susceptible to addiction.
- Additionally, **altered DRD1 function** has been linked to **cognitive flexibility**, and an inability to adapt to changing circumstances. In the context of addiction, this can result in **rigid patterns of reward-seeking behavior**, making it harder for individuals to break free from compulsive use or behavior.

The Role of DRD2 in Addiction

- **DRD2** receptors are central to the **reinforcement** of addictive behaviors. **Dopamine binding to DRD2 receptors** in the **nucleus accumbens**, the brain's central reward center, produces feelings of pleasure and satisfaction. When an individual engages in substance use or a rewarding behavior, DRD2 receptors help to reinforce the behavior by encouraging repetition.
- **Reduced DRD2 receptor density** in the brain has been observed in individuals with addiction, particularly in those who are addicted to **substances** such as cocaine or alcohol. This reduction in receptor availability means that individuals may require higher doses or more frequent engagement in the addictive behavior to experience the same level of pleasure, contributing to the cycle of **tolerance** and **dependence**.
- The dysfunction of DRD2 in addiction is also linked to **poor impulse control** and **difficulty in regulating emotions**, which can make it harder for individuals to resist cravings or overcome the urge to relapse.

The Neurobiological Mechanisms Behind Addiction

Addiction is not simply a matter of poor decision-making or lack of willpower; it is a **neurobiological disorder** deeply rooted in the brain's reward system. As the brain adapts to repeated exposure to addictive substances or behaviors, it undergoes structural and functional changes. The **dopamine system** and its receptors, particularly **DRD1** and **DRD2**, play a central role in these processes.

Neuroplasticity and Addiction

- The brain is constantly changing in response to experiences and stimuli, a phenomenon known as **neuroplasticity**. In addiction, the brain becomes increasingly conditioned to seek out the addictive behavior or substance, leading to altered wiring in the brain's reward system.
- **Over time, the brain's dopamine system becomes less responsive to natural rewards**, making the individual more reliant on the addictive substance or behavior to achieve a sense of pleasure or satisfaction. This is why addiction often leads to **diminished pleasure** in everyday activities and a growing need for more intense or frequent engagement with the addictive stimulus.
- **Changes in DRD1 and DRD2 receptor density** are a key component of this process. In some cases, **reduced DRD2 receptor density** may make it harder for the brain to register normal rewards, while **increased DRD1 receptor sensitivity** may heighten the drive to seek rewards at any cost.

Dopamine, Cravings, and Relapse

- One of the hallmark features of addiction is the presence of intense **cravings**, or the overwhelming desire to engage in the addictive behavior or use the addictive substance. These cravings are driven by the dopamine system, particularly through the activation of **DRD1** and **DRD2** receptors.
- **Cues** or triggers, such as a specific location, a social situation, or even a thought related to the addictive substance, can prompt the release of dopamine in anticipation of the reward. This anticipation reinforces the desire to use the substance or engage in the behavior, even if the individual knows it is harmful.
- Relapse is common in addiction, and it is often fueled by the dopamine-driven craving mechanism. **Changes in DRD1 and DRD2 function** can make it more difficult to manage these cravings, as the brain is primed to seek the reward despite the negative consequences.

Strategies to Balance Dopamine for Addiction Recovery

While addiction hijacks the dopamine system, recovery is possible through interventions that help **restore balance** and **retrain the brain**. By understanding the roles of **DRD1** and **DRD2**, we can better tailor strategies to optimize dopamine function and break the cycle of addiction.

Pharmacological Interventions

- Medications that target dopamine receptors are often used in addiction treatment. For example, **dopamine antagonists** (such as **antipsychotics**) can block **DRD2** receptors, helping to reduce cravings and the reinforcing effects of addictive behaviors.
- **Dopamine agonists**, which stimulate dopamine receptors, may be used in some cases to restore dopamine function in individuals with reduced receptor sensitivity. However, these must be used carefully to avoid the risk of reinforcing addictive behavior.
- **Medications like bupropion and naltrexone** have been used in addiction recovery because they affect the dopamine and opioid systems, respectively, to reduce cravings and prevent relapse.

Behavioral Therapies

- **Cognitive Behavioral Therapy (CBT)** is one of the most effective treatments for addiction. By helping individuals understand and change the thought patterns and behaviors associated with addiction, CBT can alter the brain's reward pathways and help restore balance to the dopamine system.
- **Contingency Management (CM)** and **Motivational Interviewing (MI)** are other therapies that reinforce positive behavior and motivation, thereby helping to recalibrate the reward system and reduce the appeal of addictive substances or behaviors.

Lifestyle Interventions

- Regular **exercise**, particularly aerobic exercise, has been shown to boost dopamine receptor sensitivity, improve mood, and reduce cravings. Exercise also promotes the release of **endorphins**, which can help mitigate withdrawal symptoms and reduce reliance on external rewards.
- **Mindfulness** practices and **meditation** can also help regulate dopamine levels by reducing stress and enhancing emotional regulation, which is critical in managing cravings and resisting relapse.

Conclusion

Addiction is a complex and deeply ingrained condition that involves the dysregulation of the brain's **dopamine system**, particularly through **DRD1** and **DRD2** receptors. These receptors play a central role in **reward processing, motivation**, and **reinforcement**, and their dysfunction contributes to the development and persistence of addictive behaviors. Understanding the neurobiological mechanisms behind addiction allows for targeted treatments that can help restore balance to the dopamine system, break the cycle of addiction, and promote lasting recovery. Through a combination of **pharmacological interventions, behavioral therapies**, and **lifestyle changes**, individuals can reclaim control over their lives and achieve a healthier, more balanced relationship with pleasure, reward, and motivation.

Chapter 10: Dopamine and Cognitive Function – Enhancing Learning and Memory

Dopamine, the neurotransmitter most commonly associated with the brain's reward system, also plays a crucial role in cognitive function, including learning, memory, and attention. Dopamine's influence is most prominently felt through its interaction with the **DRD1** and **DRD2** receptors, which together help regulate key cognitive processes such as **memory consolidation, processing speed, focus**, and **executive functions**. This chapter explores the link between dopamine signaling and cognitive performance, emphasizing how the proper balance of **DRD1** and **DRD2** receptor activity can optimize memory and learning abilities. It will also highlight cognitive enhancement strategies that target these receptors, offering practical tools to improve both short-term and long-term cognitive function.

The Link Between Dopamine and Cognitive Performance

Dopamine's primary role in the brain involves the modulation of **reward, motivation**, and **pleasure**, but its influence extends far beyond that. Dopamine also acts as a **cognitive enhancer**, helping to optimize attention, memory, and problem-solving capabilities. When dopamine levels are in a healthy range, individuals experience improved **mental clarity**, enhanced **focus**, and greater **mental flexibility**—all of which are essential for effective learning and memory consolidation.

The influence of dopamine on cognition is primarily mediated by two distinct but interconnected receptor systems: **DRD1** and **DRD2**.

DRD1 and Cognitive Function

- **DRD1** is particularly important in **executive functions**, which include tasks such as **planning, decision-making**, and **cognitive flexibility**. This receptor is heavily involved in processes that require individuals to adjust their thinking, initiate goal-directed behavior, and adapt strategies based on new information.
- Activation of **DRD1** also plays a role in **working memory**, the short-term storage system that allows individuals to hold and manipulate information over brief periods. Working memory is essential for problem-solving, comprehension, and reasoning, as it helps us integrate new information with previously learned material.
- The activation of **DRD1** receptors promotes **neuroplasticity**, the brain's ability to form new connections and adapt to learning experiences. This process is central to **long-term learning** and the consolidation of memories, which we will discuss in more detail later.

DRD2 and Cognitive Function

- While **DRD1** influences more complex cognitive processes, **DRD2** plays a vital role in **reward-driven learning** and **habit formation**. The **DRD2 receptor** is particularly involved in reinforcing behaviors that lead to positive outcomes, which supports learning by promoting repetition of rewarding actions.
- **DRD2** is also implicated in the ability to **process and filter information**, particularly when it comes to **attention** and **focus**. When **DRD2** is functioning optimally, individuals tend to be better at filtering out irrelevant stimuli, maintaining focus on important tasks, and avoiding distractions.
- Moreover, **DRD2** has a significant role in **motor function and coordination**, and its impact extends beyond simple cognitive tasks. By facilitating communication between the brain's reward systems and motor planning areas, **DRD2** helps ensure that behaviors are not only cognitively driven but also executed effectively.

Dopamine and Memory: From Consolidation to Recall

Memory is one of the most crucial cognitive functions influenced by dopamine. The processes of **memory consolidation** and **memory recall** are directly affected by dopamine signaling in specific brain regions, particularly the **hippocampus, prefrontal cortex**, and **striatum**.

Memory Consolidation

- **Dopamine** plays a key role in converting short-term memories into long-term ones, a process known as **memory consolidation**. The hippocampus, which is responsible for encoding new memories, relies on **dopamine signaling** to reinforce the importance of certain experiences, making them easier to remember.
- Activation of **DRD1** receptors in the hippocampus enhances the ability to consolidate and retain memories. For instance, when dopamine is released during learning or problem-solving, it signals to the brain that the information being processed is important, and should therefore be stored for future use.
- **DRD2**, on the other hand, helps to regulate how rewards and reinforcement influence memory. The presence of **dopamine** in the **striatum** reinforces the encoding of behaviors or experiences that led to positive outcomes, thus promoting memory consolidation related to pleasurable events or successful actions.

Memory Recall and Processing Speed

- Dopamine also plays a role in **memory retrieval** or recall. Research suggests that **DRD1 activation** can enhance the **accuracy** and **speed** with which memories are retrieved. This process is critical in real-time situations, where one needs to access information quickly and accurately.
- **DRD2** influences the **processing speed** of this retrieval. Individuals with higher levels of **DRD2 activity** tend to have faster access to stored memories, leading to quicker decision-making and more efficient problem-solving.
- Furthermore, **dopamine's role in attention and focus**, which is influenced by **DRD2**, enhances the brain's ability to sift through stored memories efficiently, making recall both faster and more accurate.

Cognitive Enhancement Strategies Targeting DRD1 and DRD2

Given the vital role that **DRD1** and **DRD2** play in learning, memory, and cognitive performance, optimizing their function can have a significant impact on intellectual capabilities. The following strategies can help maximize the effectiveness of these receptors and enhance cognitive performance.

Cognitive Training and Mental Exercises

- Cognitive exercises, such as **memory tasks**, **problem-solving challenges**, and **puzzles**, are one of the most effective ways to enhance dopamine signaling in the brain. These activities engage the **prefrontal cortex**, where **DRD1 receptors** are most active, and promote **neuroplasticity**.
- Studies have shown that tasks that require switching between different strategies or adapting to new situations can improve the **flexibility** of DRD1 receptors, boosting overall cognitive function and memory performance.

Physical Exercise

- **Aerobic exercise** has been shown to increase dopamine release and improve receptor sensitivity, especially in the hippocampus and prefrontal cortex. Engaging in regular physical activity boosts both **DRD1** and **DRD2** function, which enhances focus, learning capacity, and memory consolidation.
- Exercise also increases **brain-derived neurotrophic factor (BDNF)**, which plays a crucial role in **neuroplasticity**, further optimizing the brain's ability to adapt and learn new information.

Nutrition and Dopamine Modulation

- **Tyrosine**, an amino acid precursor to dopamine, plays a critical role in dopamine synthesis. Consuming foods rich in **tyrosine**, such as **lean proteins** (e.g., turkey, chicken, eggs), **fish**, **tofu**, and **dairy products**, can support dopamine production and receptor activation.
- **Omega-3 fatty acids**, found in foods like **salmon**, **walnuts**, and **flaxseeds**, have been shown to promote healthy dopamine signaling, especially in areas of the brain associated with learning and memory.

Sleep and Dopamine Regulation

- Adequate and high-quality sleep is essential for memory consolidation. During deep sleep, the brain clears out metabolic waste, strengthens synaptic connections, and consolidates newly learned information. This process is heavily influenced by **dopamine activity**, particularly in regions like the **hippocampus**.
- Disruptions in sleep can impair **dopamine signaling**, leading to memory difficulties, poor concentration, and slow processing speeds. Ensuring 7–9 hours of sleep each night will help to optimize the brain's ability to consolidate new memories and retrieve them when needed.

Mindfulness and Meditation

- **Mindfulness** practices and **meditation** have been shown to increase dopamine receptor density, particularly **DRD1**, enhancing the brain's ability to focus and process information more efficiently.
- Regular meditation can also reduce **stress**, which is often detrimental to cognitive function and memory performance. By promoting relaxation and emotional regulation, meditation helps balance dopamine levels and optimize cognitive processing.

Conclusion

Dopamine is a central player in the brain's cognitive functions, including learning, memory, and decision-making. Through the interaction of **DRD1** and **DRD2** receptors, dopamine regulates crucial cognitive processes such as **memory consolidation**, **attention, problem-solving**, and **processing speed**. By understanding and optimizing the function of these receptors, individuals can enhance their cognitive abilities, retain and recall information more effectively, and improve their overall mental performance. Practical strategies—such as **cognitive training, physical exercise, nutrition, quality sleep**, and **mindfulness practices**—offer powerful tools for boosting dopamine signaling and unlocking the brain's full potential for learning and memory. Mastering dopamine function not only improves cognitive performance but also contributes to greater mental clarity, focus, and success in everyday life.

Chapter 11: The Role of DRD1 and DRD2 in Aging and Neurodegenerative Diseases

As we age, the brain undergoes significant changes that can impact cognitive functions, emotional regulation, and motor skills. One of the key factors influencing these changes is the gradual decline in dopamine production and receptor function, particularly through the **DRD1** and **DRD2** pathways. This chapter explores how the aging process affects dopamine signaling, the implications of dopamine dysregulation in age-related diseases like **Alzheimer's** and **Parkinson's**, and the strategies that may help preserve dopamine function as we grow older.

Dopamine and Aging: The Decline in Cognitive and Motor Functions

Dopamine plays a central role in maintaining cognitive flexibility, motivation, and mood stability. As the brain ages, dopamine-producing neurons in key regions like the **striatum**, **prefrontal cortex**, and **hippocampus** naturally decline, leading to a reduction in dopamine signaling. This reduction can have wide-reaching effects on both cognitive and motor functions.

Cognitive Decline

- As dopamine activity decreases with age, there is a direct impact on **memory, learning, decision-making,** and **executive functions**. The **prefrontal cortex**, where **DRD1 receptors** are highly concentrated, is particularly vulnerable to dopamine loss, which contributes to difficulty with tasks that require planning, reasoning, and mental flexibility.
- **DRD1 receptors** are essential for **working memory** and **cognitive flexibility**, which are critical for processing and adapting to new information. A decline in **DRD1 sensitivity** or expression can contribute to **slower processing speeds** and **impaired decision-making**, common signs of aging.

Motor Decline

- The **dopamine system** is also essential for regulating **movement** through its influence on the **basal ganglia**. **DRD2 receptors** in this region help modulate the fine motor control needed for smooth, coordinated movement. As dopamine decreases with age, **motor coordination** can become impaired, contributing to symptoms such as tremors, rigidity, and slowness in movement.
- In neurodegenerative diseases like **Parkinson's disease**, the **degeneration of dopamine-producing neurons** leads to a severe decline in **DRD2 receptor function**, causing motor symptoms that significantly impact quality of life.

The Impact of DRD1 and DRD2 on Alzheimer's Disease

Alzheimer's disease (AD), one of the most common forms of dementia, is characterized by **progressive cognitive decline, memory loss**, and the inability to perform daily activities. While the primary pathology of Alzheimer's revolves around the accumulation of **amyloid plaques** and **tau tangles**, disruptions to dopamine signaling also play a significant role in the cognitive and behavioral symptoms of the disease.

DRD1 and Alzheimer's Disease

- Research has shown that dopamine dysregulation, particularly involving **DRD1 receptors**, contributes to the **cognitive impairment** observed in Alzheimer's patients. The loss of dopamine in the **prefrontal cortex**, which is crucial for memory, attention, and decision-making, can exacerbate symptoms such as **confusion, disorientation**, and **impaired learning**.
- There is also evidence suggesting that **genetic variations in DRD1** may influence an individual's susceptibility to Alzheimer's. For example, individuals with certain polymorphisms in the DRD1 gene may experience accelerated cognitive decline or an earlier onset of the disease.

DRD2 and Alzheimer's Disease

- The role of **DRD2** in Alzheimer's disease is also significant, particularly in relation to **reward processing** and **emotion regulation**. Decreased **dopamine signaling** through **DRD2 receptors** in the **ventral striatum** and **prefrontal cortex** may lead to **depression, apathy**, and **anhedonia**—all of which are common in Alzheimer's patients.
- Furthermore, **DRD2 dysfunction** contributes to the **cognitive decline** observed in Alzheimer's by impairing the brain's ability to adapt to new information or perform tasks that require executive control. As the disease progresses, **DRD2 activity** diminishes, making it more difficult for patients to initiate or sustain complex thought processes.

The Impact of DRD1 and DRD2 on Parkinson's Disease

Parkinson's disease (PD) is a neurodegenerative disorder marked by **motor dysfunction**, including **tremors, rigidity, bradykinesia** (slowness of movement), and **postural instability**. These symptoms result from the loss of **dopamine-producing neurons** in the **substantia nigra**, which are critical for the **DRD1** and **DRD2 receptors** involved in motor control.

DRD1 and Parkinson's Disease

- **DRD1 receptors** in the **prefrontal cortex** and **striatum** are integral to maintaining **motor control**, **learning**, and **decision-making**. In Parkinson's disease, the **degeneration of dopamine-producing neurons** leads to a decrease in the activation of **DRD1**, which impacts the brain's ability to control movement.
- In addition to motor symptoms, **DRD1 dysfunction** in Parkinson's patients can contribute to **cognitive decline**, including difficulties with **working memory**, **planning**, and **executive functions**. As the disease progresses, individuals may experience **dementia**, which is characterized by worsened cognitive and functional abilities.

DRD2 and Parkinson's Disease

- The role of **DRD2** in Parkinson's disease is especially critical in the regulation of movement. **DRD2 dysfunction** is a hallmark of Parkinson's, as the loss of dopamine leads to impaired **motor coordination** and **reward-based learning**. This manifests in the motor symptoms commonly associated with the disease, including **tremors**, **stiffness**, and **slow movement**.
- Studies suggest that **dopamine agonists** targeting **DRD2 receptors** can help alleviate some of the motor symptoms of Parkinson's by partially restoring dopamine signaling in the **basal ganglia**. However, over time, **dopamine therapy** may become less effective due to further receptor degradation.

Approaches to Preserve Dopamine Function with Aging

Although age-related declines in dopamine function are inevitable to some extent, there are several strategies to slow down or mitigate these changes, particularly in relation to **DRD1** and **DRD2** function. These strategies focus on promoting **dopamine receptor sensitivity**, supporting **dopamine production**, and improving overall brain health.

Physical Exercise

- Regular aerobic exercise, such as **walking, cycling,** or **swimming,** has been shown to promote **dopamine release** and improve the function of **dopamine receptors**. Exercise also increases **neurogenesis**, the creation of new neurons, particularly in the **hippocampus**, where **dopamine** is critical for learning and memory.
- Studies indicate that exercise increases **DRD1 and DRD2 receptor density**, which helps preserve **cognitive function** and **motor control** as we age.

Diet and Nutrition

- Consuming a **nutrient-rich diet** that includes foods high in **tyrosine** (the amino acid precursor to dopamine) can support **dopamine synthesis. Lean proteins, fish, eggs**, and **dairy** are excellent sources of tyrosine.
- Antioxidant-rich foods like **berries, dark chocolate**, and **green tea** can help protect dopamine-producing neurons from oxidative stress, which accelerates neurodegeneration.
- Supplementing with **omega-3 fatty acids**, found in **fatty fish** like **salmon** and **mackerel**, can help improve **dopamine receptor function** and promote brain plasticity.

Mental Stimulation

- Engaging in **cognitive exercises**, such as **puzzles, reading**, and **learning new skills**, can help maintain **cognitive flexibility** and **working memory**, which are essential for healthy aging. **Mental stimulation** boosts dopamine release, particularly in areas such as the **prefrontal cortex**.
- **Mindfulness meditation** and **brain training apps** that challenge attention and memory can also enhance dopamine signaling and protect against cognitive decline.

Pharmacological Interventions

- Although still under investigation, certain **dopamine-targeting drugs** may help alleviate the symptoms of aging-related neurodegenerative diseases. For instance, **dopamine agonists** can help **Parkinson's disease** patients by activating **DRD2 receptors**, thereby improving motor symptoms.
- In Alzheimer's disease, **dopamine precursors** or **modulators** may support cognitive function by boosting **dopamine signaling** in the **prefrontal cortex** and **hippocampus**.

Conclusion

As we age, the decline in dopamine function, particularly through **DRD1** and **DRD2 receptors**, can contribute to cognitive and motor impairments commonly seen in neurodegenerative diseases like **Alzheimer's** and **Parkinson's**. However, by adopting strategies such as regular exercise, a healthy diet, mental stimulation, and potential pharmacological interventions, it is possible to **preserve dopamine function** and mitigate age-related cognitive decline. Understanding the role of dopamine in aging not only provides valuable insights into how we can manage the risks of **neurodegenerative diseases**, but it also offers hope for improving quality of life as we grow older. By nurturing our **dopamine system**, we can promote a healthier, more vibrant aging process.

Chapter 12: The Impact of Diet, Exercise, and Lifestyle on Dopamine Receptors

In the quest to optimize mental and emotional well-being, understanding how lifestyle factors like **diet, exercise**, and overall **wellness habits** impact **dopamine receptors** is essential. Dopamine, often referred to as the "feel-good" neurotransmitter, is responsible for motivation, pleasure, and reward processing. However, dopamine receptor sensitivity, specifically the **DRD1** and **DRD2 receptors**, can be modulated by what we consume, how we move, and how we manage our environment. This chapter explores the ways in which **diet, physical activity**, and **lifestyle** choices can influence dopamine signaling and receptor functionality, offering practical tips for optimizing **dopamine balance** to promote cognitive and emotional health.

The Role of Diet in Dopamine Receptor Sensitivity

Nutrition plays a pivotal role in the regulation of dopamine levels and receptor sensitivity. Certain nutrients directly influence dopamine production, while others support the **health of dopamine receptors**, ensuring efficient signaling in the brain.

Tyrosine and Phenylalanine

- **Tyrosine** is the amino acid precursor to dopamine. It is converted into **L-DOPA**, which then synthesizes dopamine in the brain. Foods rich in **tyrosine** (such as **lean meats, fish, eggs, dairy products**, and **soy products**) are key to boosting dopamine production.
- **Phenylalanine**, found in high-protein foods like **meats, cheese, seeds**, and **nuts**, is converted into tyrosine, further supporting dopamine synthesis.

Omega-3 Fatty Acids

- Omega-3 fatty acids, particularly **DHA** (docosahexaenoic acid), are essential for maintaining the structural integrity of brain cells, including the receptors involved in dopamine signaling. Foods rich in omega-3s, such as **fatty fish** (e.g., **salmon, mackerel, sardines**) and **flaxseeds, chia seeds**, and **walnuts**, help preserve dopamine receptor function, particularly **DRD1** and **DRD2**.
- Omega-3s have also been shown to improve **neuroplasticity**, allowing the brain to adapt more efficiently to new experiences, which is vital for **learning, memory**, and **emotional regulation**.

Antioxidants

- The brain is highly susceptible to oxidative stress, which can damage dopamine-producing neurons and receptors. Antioxidants, such as **vitamin C, vitamin E**, and **polyphenols**, protect dopamine receptors from this oxidative damage.
- Foods rich in antioxidants, like **berries, dark chocolate, green tea**, and **spinach**, help neutralize harmful free radicals and preserve the integrity of dopamine function.

B-Vitamins

B-vitamins, especially

,

, and

, are crucial for dopamine production and receptor sensitivity.

, in particular, is involved in the conversion of

to dopamine. Sources of B-vitamins include

,

,

, and

.

Magnesium and Zinc

- **Magnesium** helps regulate dopamine release and supports the **functionality of dopamine receptors**, particularly in the **prefrontal cortex**, an area involved in executive function and emotional regulation. Magnesium-rich foods include **spinach, almonds, avocados**, and **bananas**.
- **Zinc** also plays a role in maintaining dopamine receptor sensitivity, and deficiencies in zinc have been linked to dopamine dysregulation. Foods rich in zinc include **pumpkin seeds, cashews**, and **chickpeas**.

Physical Exercise and Its Influence on Dopamine Receptors

Exercise has profound effects on the brain's dopamine system. Regular physical activity not only boosts dopamine levels but also enhances **dopamine receptor sensitivity**, which is essential for motivation, learning, and mood regulation.

Aerobic Exercise

- Aerobic activities like **running**, **swimming**, and **cycling** have been shown to increase dopamine production in areas such as the **striatum** and **prefrontal cortex**. This has a direct impact on motivation, mood, and cognitive function.
- Aerobic exercise also increases the **density of dopamine receptors**, particularly **DRD1** and **DRD2**, in key areas of the brain involved in learning and behavior change. This enhancement of receptor function can improve decision-making, emotional regulation, and overall brain plasticity.

Resistance Training

- Strength training, or **resistance exercise**, has also been found to enhance dopamine receptor sensitivity, especially in the **basal ganglia**, which is responsible for motor control and reward processing.
- Research suggests that resistance training may improve **dopamine signaling** related to motivation and goal-directed behavior, helping to combat the feelings of apathy and lack of drive often associated with age or mental health conditions like depression.

Endorphin Release

Physical activity stimulates the release of

, which interact with dopamine pathways to enhance feelings of well-being. The release of endorphins during exercise can increase

, leading to improved mood and reduced stress.

Neurogenesis and Neuroplasticity

Exercise also promotes

, the creation of new neurons, particularly in the

, where dopamine plays a role in learning and memory. By encouraging brain plasticity, exercise supports long-term cognitive function and the preservation of dopamine signaling pathways.

Lifestyle Modifications to Optimize Dopamine Balance

Beyond diet and exercise, other lifestyle factors can influence dopamine receptor function and overall brain health. These include sleep, stress management, and social interactions, all of which play a critical role in dopamine regulation.

Sleep and Dopamine Regulation

- Quality sleep is essential for optimal dopamine function. During sleep, the brain undergoes processes of repair and recalibration, including the restoration of **dopamine receptor sensitivity**.
- Poor sleep or chronic sleep deprivation can lead to **dopamine receptor downregulation**, making it harder to experience pleasure and motivation during waking hours. Consistently getting **7-9 hours of restorative sleep** each night helps maintain a healthy dopamine system, supports memory consolidation, and improves emotional regulation.

Stress Management

- Chronic stress is a major disruptor of dopamine function. High levels of stress hormones like **cortisol** can damage dopamine receptors, especially in regions of the brain that regulate mood and motivation, such as the **prefrontal cortex**.
- Techniques such as **mindfulness meditation, deep breathing exercises, yoga**, and **progressive muscle relaxation** help reduce stress and protect dopamine pathways. These practices promote **dopamine receptor sensitivity**, enhance emotional balance, and support cognitive clarity.

Social Connection

- Positive social interactions have been shown to trigger the release of dopamine, particularly in the brain's **reward pathways**. Healthy social relationships help regulate mood and provide a sense of purpose and motivation, which can enhance **dopamine receptor activity**.
- Engaging in **meaningful social activities** like spending time with loved ones, participating in community events, or joining group hobbies not only improves emotional well-being but also optimizes dopamine signaling by reinforcing the brain's reward system.

Practical Tips for Dopamine Optimization Through Lifestyle Changes

To optimize dopamine balance and receptor function, consider adopting these practical strategies:

1. **Incorporate dopamine-boosting foods** into your diet. Focus on **protein-rich foods, omega-3 fatty acids**, and **antioxidant-rich vegetables and fruits.**
2. **Exercise regularly**—combine aerobic exercise with strength training for maximum benefit to your dopamine receptors.
3. Prioritize **sleep** by following a regular sleep schedule and creating a **sleep-friendly environment**.
4. Practice **stress management techniques** such as **mindfulness** or **yoga** to protect your dopamine system from the effects of chronic stress.
5. Cultivate **positive social connections** and engage in rewarding social activities that promote dopamine release.
6. Consider integrating **brain-boosting supplements** like **magnesium, zinc, and B-vitamins**, which support dopamine production and receptor function.

Conclusion

By making intentional choices about **diet**, **exercise**, and **lifestyle**, individuals can enhance the sensitivity and functionality of **dopamine receptors** in the brain. **DRD1** and **DRD2**, the key players in motivation, emotional regulation, and learning, are sensitive to environmental factors that can either optimize or hinder their effectiveness. By understanding the profound impact that lifestyle changes can have on dopamine signaling, individuals can create a personalized plan for improving mental clarity, motivation, mood, and overall cognitive health. This holistic approach to dopamine balance provides a foundation for a healthier, more fulfilling life.

Chapter 13: Pharmacological Interventions – Medications Targeting DRD1 and DRD2

Pharmacological interventions targeting dopamine receptors—specifically **DRD1** and **DRD2**—play a significant role in managing a range of cognitive and emotional conditions. From treating mental health disorders to enhancing performance, medications that modulate dopamine activity are pivotal in achieving therapeutic goals. This chapter explores the current landscape of pharmacological treatments designed to affect dopamine receptor function, how these medications work, and the future directions in pharmaceutical research aimed at optimizing dopamine balance.

Overview of Pharmacological Treatments Targeting Dopamine Receptors

Medications that influence **dopamine receptors** are used to treat a variety of psychiatric and neurological disorders, including **schizophrenia, bipolar disorder, Parkinson's disease, addiction**, and **depression**. By altering dopamine signaling, these treatments can either increase or decrease dopamine activity, depending on the therapeutic need.

Antipsychotics (Dopamine Antagonists)

- **Antipsychotic drugs** are commonly prescribed for conditions like schizophrenia, bipolar disorder, and severe depression. These medications primarily act by blocking dopamine receptors, especially **DRD2**, in brain regions associated with psychosis and emotional regulation.
- The **first-generation antipsychotics** (e.g., **haloperidol** and **chlorpromazine**) were the first to target dopamine, specifically **DRD2**, to reduce symptoms like delusions and hallucinations. These medications can lead to side effects like **extrapyramidal symptoms** (motor control problems) due to over-blockade of dopamine receptors.
- **Second-generation antipsychotics** (e.g., **risperidone**, **olanzapine**, and **quetiapine**) are more selective and often have fewer motor side effects. These newer drugs also affect **DRD1** and **DRD2**, as well as other neurotransmitter systems, such as serotonin, to improve mood and cognitive function while reducing psychotic symptoms.

Stimulants (Dopamine Agonists)

- **Stimulant medications**, such as **methylphenidate** (e.g., **Ritalin**) and **amphetamine-based** drugs (e.g., **Adderall**), are used primarily to treat **attention-deficit hyperactivity disorder (ADHD)**. These drugs enhance dopamine transmission by increasing the release of dopamine into synaptic spaces, specifically acting on **DRD1** and **DRD2** receptors.
- By increasing dopamine availability in the prefrontal cortex, stimulants help improve focus, motivation, and cognitive control, making them effective for individuals with ADHD or cognitive disorders characterized by low dopamine activity.

Dopamine Agonists (For Parkinson's Disease)

- **Dopamine agonists**, such as **pramipexole**, **ropinirole**, and **bromocriptine**, are used to treat **Parkinson's disease**, a neurodegenerative condition associated with the depletion of dopamine-producing neurons. These drugs directly stimulate dopamine receptors, primarily **DRD2**, to improve motor control and reduce symptoms like rigidity, tremors, and bradykinesia (slowness of movement).
- While these drugs can significantly improve symptoms, they come with potential side effects, including **impulse control disorders**, where patients may experience compulsive behaviors (e.g., gambling, hypersexuality) due to overstimulation of the dopamine system.

Antidepressants

- Certain **antidepressant medications** (e.g., **bupropion** and **tricyclic antidepressants**) also affect dopamine levels. **Bupropion** works by inhibiting the reuptake of dopamine and norepinephrine, thereby increasing the availability of dopamine at synapses, especially in areas like the **prefrontal cortex** and **nucleus accumbens**.
- This mechanism of action helps alleviate symptoms of **depression**, **anxiety**, and **fatigue**, especially in individuals whose dopamine systems may be underactive.

Opioid Antagonists and Addiction Treatments

- Medications such as **naltrexone** and **buprenorphine** are used to treat opioid addiction by blocking the rewarding effects of drugs like heroin and morphine. These drugs primarily interact with opioid receptors, but they also influence dopamine release through interactions with the **mesolimbic** dopamine pathway.
- By reducing the dopamine "reward" from drug use, these medications help reduce cravings and prevent relapse in individuals recovering from addiction. Some **opioid antagonists** also act on **DRD2** and other receptors to reduce the intensity of addictive behavior.

How Medications Affect DRD1 and DRD2

DRD1

Cognitive Enhancement

- Research into the modulation of **DRD1** suggests that drugs targeting this receptor may help improve **cognitive flexibility, executive function**, and **memory**. For example, some **dopamine agonists** and **psychostimulants** increase **DRD1** activity, which can enhance motivation and learning.
- In patients with cognitive disorders, such as **Parkinson's disease** or **schizophrenia**, medications that increase **DRD1** activation may help mitigate cognitive deficits, particularly in working memory and decision-making processes.

DRD2

- **DRD2** plays a crucial role in regulating mood and pleasure-seeking behaviors. Antipsychotics that block **DRD2** are effective in treating conditions like **schizophrenia** and **bipolar disorder** by dampening excessive dopamine transmission in areas of the brain linked to **psychotic symptoms**.
- However, overstimulation of **DRD2**, such as from stimulant abuse or **dopamine agonists**, can lead to **addictive behaviors** and **impulse control issues**, as seen with medications used in Parkinson's disease treatment.

Future Directions in Dopamine-Related Pharmaceutical Development

The field of dopamine-related pharmaceuticals is evolving rapidly, with new discoveries offering hope for more precise and effective treatments for a range of conditions.

Targeted Receptor Modulation

- Rather than using broad-spectrum drugs that affect both **DRD1** and **DRD2** indiscriminately, researchers are now focusing on developing more **targeted** drugs that selectively modulate one receptor subtype. This would reduce side effects and enhance the therapeutic benefits of dopamine-related drugs.
- For instance, **partial agonists** that activate **DRD1** without overstimulating it, or **DRD2-selective antagonists** that provide therapeutic effects for conditions like schizophrenia without affecting motor function, are a promising area of research.

Gene Therapy and CRISPR

- **Gene editing technologies** like **CRISPR** offer exciting possibilities for altering the expression of dopamine receptors at the genetic level. By correcting mutations in **DRD1** and **DRD2** genes, it may be possible to prevent or reverse certain dopamine-related disorders.
- For example, genetic interventions may hold the potential to reduce the risk of **dopamine receptor deficiencies** that contribute to conditions such as **ADHD**, **Parkinson's disease**, or **schizophrenia**, potentially offering a more permanent solution than pharmacological interventions.

Personalized Medicine

- With advancements in **pharmacogenomics**, which tailors drug treatments based on an individual's genetic profile, treatments for dopamine-related conditions may become more personalized. By understanding a person's specific **dopamine receptor genes**, doctors could prescribe medications that are more likely to work effectively and minimize side effects.
- For example, patients with certain genetic variations of **DRD1** or **DRD2** may respond differently to antipsychotics, antidepressants, or stimulants. Personalized approaches could ensure that individuals receive the most effective treatment based on their unique genetic makeup.

Neurostimulation and Non-Invasive Techniques

- In addition to pharmaceutical treatments, **non-invasive brain stimulation** methods, such as **transcranial magnetic stimulation (TMS)** and **deep brain stimulation (DBS)**, are being explored for their ability to modulate dopamine activity in the brain. These techniques can directly influence dopamine circuits, offering a potential treatment for conditions like **depression, Parkinson's disease**, and **chronic pain**.
- Future research may explore how these technologies can complement or replace traditional medications, offering a more precise way to modulate dopamine receptor function.

Conclusion

Pharmacological interventions targeting dopamine receptors are integral to treating a range of disorders involving **dopamine dysregulation**, including **schizophrenia, Parkinson's disease, ADHD**, and **addiction**. By influencing **DRD1** and **DRD2** receptor function, these medications can restore balance to the brain's reward systems, improving mood, cognitive function, and behavior. However, as research into dopamine modulation continues, the future holds promise for **more selective treatments** that offer greater precision, fewer side effects, and enhanced efficacy. Advances in **genetic medicine, personalized treatments**, and **neurostimulation** hold the potential to revolutionize how we approach dopamine-related conditions, offering hope for improved mental health and cognitive function across the lifespan.

Chapter 14: Dopamine and Emotional Regulation – Mastering Mood and Motivation

Dopamine is often heralded as the brain's "feel-good" neurotransmitter, but its role extends far beyond simply promoting pleasure or reward. It is deeply intertwined with the regulation of **mood, motivation**, and **emotional balance**. Specifically, **DRD1** and **DRD2** receptors play key roles in how we experience emotions, process rewards, and react to external stimuli, making them pivotal in managing emotional states. Mastering dopamine regulation—by optimizing the function of these receptors—can help individuals achieve greater emotional stability, motivation, and a more fulfilling life.

The Connection Between Dopamine and Emotional States

Dopamine is fundamentally linked to **emotion regulation**. It influences how we perceive and react to the world around us. When dopamine levels are balanced, individuals tend to experience **positive emotions, increased motivation**, and **emotional resilience**. However, dysregulated dopamine signaling—whether through too much or too little dopamine activity—can lead to mood instability, heightened stress responses, and difficulty managing emotional reactions.

The two primary dopamine receptors, **DRD1** and **DRD2**, each contribute in unique ways to the regulation of emotional states:

DRD1

- **Motivation and Reward:** DRD1 is crucial for the **anticipation of rewards** and the **pursuit of goals**. Its activation is associated with **positive reinforcement**—when an individual successfully works toward a goal or experiences success, dopamine release enhances feelings of reward and reinforces the behavior.
- **Mood and Emotional Flexibility:** DRD1 also impacts emotional **flexibility**, enabling individuals to shift and adapt to changing emotional landscapes. An optimal DRD1 function promotes the ability to regulate emotions and cope with stress, making it easier to maintain a balanced emotional state in the face of challenges.

DRD2

- **Pleasure and Satisfaction:** DRD2 plays a significant role in **pleasure-seeking behaviors** and the experience of **satisfaction**. It is implicated in the **brain's reward system**, affecting our sense of pleasure and emotional fulfillment. Low DRD2 activity is often linked with feelings of emotional numbness or anhedonia (inability to feel pleasure), whereas overactive DRD2 signaling can contribute to excessive cravings and impulsivity.

- **Habit Formation and Emotional Conditioning:** DRD2's influence on **habit formation** and **emotional conditioning** means that it can also affect emotional responses to past experiences. Its activation can solidify emotional patterns, both positive and negative, influencing how we react to similar situations in the future.

Techniques to Regulate Dopamine to Foster Positive Emotions and Mood Stability

Given the critical role dopamine plays in mood regulation, there are several techniques to optimize **dopamine receptor activity** and maintain emotional stability. These strategies are aimed at balancing the activity of **DRD1** and **DRD2** to foster positive emotional states and boost motivation.

Mindful Goal-Setting and Achievement

- **Structured goal-setting** stimulates **DRD1** by fostering a sense of purpose and accomplishment. When you set clear, achievable goals and make steady progress toward them, your brain releases dopamine in anticipation of the reward, thereby reinforcing positive behaviors and emotional regulation.
- Focus on **small wins**: Breaking large tasks into smaller, manageable goals can provide frequent opportunities for dopamine release, leading to a sustained sense of motivation and fulfillment.

Balanced Reward Systems

- Regularly engage in **rewarding activities** that align with your values and long-term goals. These can include hobbies, physical activities, or creative endeavors that enhance **intrinsic motivation**. Balancing **intrinsic** rewards (which are internal and self-driven) with **extrinsic** rewards (such as recognition or material gain) helps keep dopamine levels stable without overstimulating **DRD2**, which can lead to dependence or impulsive behavior.
- **Delayed gratification** is another powerful strategy. By training yourself to delay immediate rewards in favor of long-term benefits, you can better regulate **DRD2** activity and enhance your ability to manage emotional responses.

Physical Activity

- Exercise has been shown to significantly boost dopamine levels, enhancing both **DRD1** and **DRD2** receptor function. Activities like **aerobic exercise, strength training**, or **yoga** have been found to improve mood and emotional resilience. Regular physical activity can help maintain a **healthy dopamine balance**, reducing stress and preventing mood swings.
- **Endorphin release** during physical exercise not only enhances dopamine levels but also activates the brain's **reward centers**, improving emotional states over time.

Dietary Choices

- **Nutritional intake** can directly impact dopamine receptor sensitivity and activity. Eating foods rich in **tyrosine**—an amino acid precursor to dopamine—can support the production and release of dopamine. Foods like **bananas, avocados, eggs**, and **fish** are excellent sources.
- **Antioxidants**, found in fruits and vegetables, also help protect dopamine receptors from oxidative stress, preserving their function and enhancing emotional regulation.

Sleep and Stress Management

- A good night's sleep is essential for maintaining **dopamine receptor sensitivity**. **Chronic sleep deprivation** can impair dopamine receptor function, particularly **DRD2**, making it harder to experience pleasure and motivation. Ensuring **7-9 hours of restful sleep** per night can help optimize dopamine balance.
- Reducing **chronic stress** is equally important, as prolonged stress can lead to **dopamine depletion** and **decreased receptor sensitivity**. Techniques such as **mindfulness meditation, deep breathing exercises**, and **progressive muscle relaxation** can help lower cortisol levels, protect dopamine function, and stabilize emotional responses.

Cognitive-Behavioral Strategies for Balancing Dopamine Receptors

Cognitive-behavioral therapy (CBT) and other therapeutic strategies can directly influence dopamine activity by promoting healthier thought patterns and emotional responses. Here are some techniques that leverage the brain's reward systems for emotional regulation:

Cognitive Restructuring

CBT techniques like

help individuals identify and challenge negative thought patterns. By recognizing automatic thoughts that fuel anxiety or depression, individuals can replace them with more balanced, realistic beliefs. This shift reduces negative emotional responses and helps regulate

, particularly through

, by fostering positive reinforcement and goal-directed behavior.

Behavioral Activation

This technique involves increasing engagement in

that lead to positive emotional experiences. By strategically scheduling activities that trigger the brain's reward system, individuals can boost

activity in a controlled manner, improving mood and decreasing feelings of anhedonia.

Gratitude Practices

Practicing

has been shown to enhance emotional well-being by fostering a positive feedback loop in the brain. Focusing on what is good in your life can trigger the release of dopamine, supporting both

and

activation. A simple daily gratitude practice—writing down things you're thankful for—can help maintain emotional balance and promote happiness.

Conclusion

The mastery of **dopamine regulation** is a powerful tool for managing emotional states, boosting motivation, and achieving personal well-being. By understanding the roles of **DRD1** and **DRD2**, individuals can better navigate the complexities of mood, motivation, and emotional regulation. Whether through mindful goal-setting, physical activity, balanced rewards, or cognitive-behavioral strategies, there are many practical approaches to enhance dopamine function and achieve emotional stability. Ultimately, the key to emotional mastery lies in maintaining the delicate balance between the brain's reward systems, ensuring that dopamine levels are neither too high nor too low. By mastering this balance, individuals can unlock their true potential for **optimal emotional health** and **motivated living**.

Chapter 15: Harnessing Dopamine for Peak Performance – Achieving Optimal Mental and Physical States

In the pursuit of excellence, whether in the realm of professional achievements, athletic prowess, or cognitive performance, **dopamine** plays a central role in propelling us forward. By understanding and optimizing **dopamine receptors**, particularly **DRD1** and **DRD2**, individuals can unlock their full potential for peak performance, sustained motivation, and elevated mental clarity. This chapter explores how dopamine influences **performance**, how high achievers leverage dopamine for success, and actionable strategies to harness dopamine to reach optimal mental and physical states.

The Role of Dopamine in Performance

Dopamine is intimately tied to **motivation, reward** systems, and the **brain's executive functions**. It is the neurotransmitter responsible for propelling us toward goals, rewarding our progress, and helping us maintain focus and energy. Dopamine's influence on **cognitive function**—such as **decision-making, learning, attention**, and **memory**—makes it essential not only for mental clarity but also for **physical performance** in sports and other high-demand activities.

Two specific dopamine receptors, **DRD1** and **DRD2**, are crucial for regulating these performance-related processes:

DRD1

- DRD1 plays a pivotal role in goal-directed behavior, including **planning** and **executing actions** necessary to achieve objectives. This receptor is associated with the **anticipation of rewards** and the **drive to overcome obstacles**. DRD1 activation strengthens cognitive functions like **executive decision-making** and **problem-solving**, making it a cornerstone for peak performance.
- Optimal DRD1 activity supports **focus, cognitive flexibility**, and the ability to adapt in the face of setbacks—key qualities in both intellectual and physical endeavors.

DRD2

- DRD2 governs the **reward** and **pleasure** systems in the brain, influencing how individuals respond to the pursuit of **immediate and long-term rewards**. In performance contexts, DRD2 regulates how rewarding the process feels—how much satisfaction is derived from progress or success.
- When DRD2 is activated in balance with DRD1, it promotes sustained engagement without causing burnout or overindulgence in rewards. It ensures that individuals experience pleasure and satisfaction while avoiding addictive behaviors or excessive risk-taking.

The Psychology of Motivation and Flow States

One of the most powerful ways dopamine impacts performance is through the **psychological state of flow**. Flow is that elusive state in which an individual is so fully immersed in a task that they lose track of time and perform at their highest potential. Achieving flow is a powerful sign of **dopamine's optimal function** in the brain, as it enhances **focus**, **creativity**, and **problem-solving abilities**.

Key elements of flow include:

- **Clear goals**: Knowing what needs to be accomplished and having a sense of progression provides continuous dopamine rewards as individuals work toward completion.
- **Challenge-skill balance**: Flow occurs when a task is appropriately challenging, pushing one's skills to their limits but not so difficult as to cause frustration or failure. This balance ensures **dopamine release** throughout the experience, keeping the individual in a motivated, engaged state.
- **Immediate feedback**: Feedback, especially positive reinforcement, triggers dopamine release and signals to the brain that the goal is within reach, fueling the drive to continue.
- **Total concentration**: Flow requires complete **attention** and **immersion**, which enhances dopamine levels by limiting distractions and ensuring full engagement.

By understanding how to trigger and sustain dopamine-driven flow states, individuals can optimize their cognitive and physical performance.

Tools to Optimize Dopamine Balance for Sustained Productivity

To achieve and maintain peak performance over time, it is essential to maintain an optimal balance of **dopamine signaling** through various strategies. Overactivation or underactivation of dopamine can lead to burnout, lack of motivation, or diminished cognitive abilities. Below are tools that can help fine-tune dopamine levels to promote sustained high performance.

Structured Goal-Setting and Visualization

- Dopamine is strongly influenced by goal-directed behavior. The process of setting clear, **achievable goals** creates dopamine-driven motivation, propelling individuals to complete tasks and pursue further success.
- **Visualization** is a potent tool for enhancing motivation. By vividly imagining the process and the rewards, you strengthen the neural pathways related to DRD1 activation, which supports goal-oriented behavior. Visualizing success also triggers **dopamine release** in anticipation of future rewards.

Exercise and Physical Activity

- Exercise is a well-documented method to boost dopamine levels, especially through its effect on **DRD1** and **DRD2** receptors. High-intensity interval training (HIIT) and strength training have been shown to increase **dopamine receptor density**, improving mood and cognitive function, both of which are essential for peak performance.
- Physical activity not only provides a release of dopamine but also enhances **neuroplasticity**, the brain's ability to adapt and optimize performance over time.

Microdosing and Nootropics

- The use of **nootropics**—substances that enhance cognitive function—has gained popularity among high-performers. Some nootropics directly influence dopamine signaling, particularly those that increase dopamine receptor sensitivity. Substances like **L-theanine, methylphenidate**, and certain **adaptogens** (e.g., Rhodiola rosea) can balance dopamine and improve focus, memory, and cognitive function.
- Another emerging practice is **microdosing**, the practice of taking small, sub-perceptual doses of psychedelics (such as **psilocybin** or **LSD**). Some research suggests that microdosing may increase dopamine release in a controlled manner, improving mood, creativity, and focus.

Stress Management and Recovery

- Chronic stress can result in **dopamine depletion** and **downregulation of receptors**, leading to diminished performance. **Stress management techniques** such as deep breathing, progressive muscle relaxation, and yoga help maintain dopamine levels while reducing the negative effects of stress on cognitive function.
- Adequate **rest and recovery** are also vital. **Sleep**, in particular, allows the brain to restore dopamine levels and optimize cognitive performance for the following day. Ensuring **7-9 hours of sleep** each night enhances **dopamine receptor sensitivity**, preparing the brain for peak performance the next day.

Diet and Nutrition

- **Nutritional support** plays a significant role in optimizing dopamine function. Foods high in **tyrosine**, such as lean proteins (chicken, fish, eggs), **nuts**, and **seeds**, serve as precursors for dopamine synthesis. Incorporating these into your diet can promote sustained dopamine production, which supports motivation and cognitive performance.
- **Omega-3 fatty acids**, found in fatty fish (like salmon) and walnuts, support brain health and the function of dopamine receptors, enhancing cognitive performance and emotional regulation.

Achieving Peak Physical Performance with Dopamine

For athletes and individuals engaged in physical training, dopamine's impact on performance is equally profound. **DRD1** helps athletes stay motivated and focused during training, while **DRD2** governs the feeling of satisfaction derived from completing physical challenges.

To optimize **dopamine function for athletic performance**, consider these strategies:

1. **Optimized Training Cycles**: Similar to the psychological concept of "flow," athletes can optimize training cycles to create the right balance of **challenge and recovery**. Training programs that progressively increase difficulty, combined with periods of rest, help maintain **dopamine sensitivity** while promoting growth and improvement.
2. **Reward-Based Systems**: Creating a reward-based system for training milestones—whether through tangible rewards or personal milestones—activates **dopamine release**, keeping athletes motivated and focused on their long-term goals.
3. **Endorphin Release**: Regular endurance activities, such as running or cycling, can trigger the release of **endorphins**, which in turn boost dopamine levels. This synergy supports both mental and physical resilience.

Conclusion

Mastering **dopamine regulation** is essential for achieving **peak performance**, whether in the intellectual, emotional, or physical realms. By understanding how **DRD1** and **DRD2** receptors influence motivation, reward, and emotional states, individuals can optimize their cognitive and physical capacities. Whether through strategic goal-setting, exercise, stress management, or nutritional support, the tools for mastering dopamine balance are within reach. By harnessing the power of dopamine, anyone can achieve sustained high performance, navigate challenges with resilience, and continually push toward greater success.

Chapter 16: Mindfulness and Meditation – Balancing Dopamine through Mental Practices

In the quest to optimize **dopamine function**, **mindfulness** and **meditation** have emerged as powerful tools. These practices not only promote **mental clarity** and emotional stability but also have a direct impact on **dopamine receptor function**, particularly **DRD1** and **DRD2**. In this chapter, we explore how these ancient techniques can help regulate dopamine levels, enhance cognitive and emotional balance, and foster a greater sense of well-being.

The Role of Mindfulness in Regulating Dopamine Levels

Mindfulness refers to the practice of being fully present and aware in the moment, without judgment. This state of **focused attention** helps reduce stress, increase emotional regulation, and improve overall mental health. When practiced regularly, mindfulness can have profound effects on dopamine signaling in the brain.

Reduced Stress and Dopamine Balance

- Chronic stress is one of the primary contributors to **dopamine dysregulation**. It leads to the depletion of dopamine stores and can result in the downregulation of dopamine receptors, particularly **DRD2**. By engaging in mindfulness practices such as focused breathing or body scan techniques, individuals can lower stress hormones like cortisol, which in turn helps to maintain a healthy balance of dopamine.
- Studies have shown that mindfulness practices activate regions of the brain involved in **emotional regulation** (like the prefrontal cortex) and **reward processing** (like the striatum). This balance encourages healthy dopamine release, enhancing **motivation** and **pleasure** without the risk of burnout or addictive behaviors.

Enhanced Cognitive Function and DRD1 Activation

Mindfulness not only helps with emotional stability but also enhances

. By practicing mindfulness, individuals can activate

, which plays a key role in decision-making, goal-setting, and adapting to new situations. Mindfulness promotes

, making it easier to focus on tasks and resist distractions, a critical aspect of cognitive performance.

Neuroplasticity and Dopamine

Research indicates that mindfulness practice encourages

, the brain's ability to reorganize itself by forming new neural connections. By strengthening the pathways associated with dopamine receptors, mindfulness can increase the

of

and

, which may enhance

and

over time.

Meditation Techniques that Influence DRD1 and DRD2 Function

Meditation, often paired with mindfulness, is another mental practice that can have a powerful influence on **dopamine receptors**. Different forms of meditation activate distinct neural circuits, each with implications for dopamine balance.

Focused Attention Meditation

- This type of meditation involves concentrating on a single object, breath, or thought, helping to cultivate deep concentration. Research shows that focused attention meditation increases the density of **dopamine receptors**, particularly **DRD1**, in brain areas like the **prefrontal cortex**, which is involved in goal-directed behavior and executive functions.
- By regularly engaging in focused attention, individuals enhance their ability to stay focused, plan effectively, and make thoughtful decisions, all of which are linked to optimal dopamine functioning.

Loving-Kindness Meditation (LKM)

- Loving-kindness meditation is designed to foster **compassion**, **empathy**, and positive emotional states toward oneself and others. This form of meditation has been shown to increase **dopamine release** by activating the **reward centers** of the brain. This effect is partly due to the positive emotions generated by feelings of love and compassion, which stimulate dopamine and promote mental well-being.
- As LKM helps regulate emotional states, it can improve **DRD2 receptor function**, leading to greater emotional stability, better mood regulation, and a more balanced sense of pleasure.

Mindfulness-Based Stress Reduction (MBSR)

MBSR

reduce symptoms of anxiety

depression

stress

dopamine signaling

emotional resilience

mental clarity

Transcendental Meditation (TM)

TM involves the use of a mantra to facilitate deep relaxation and awareness. This technique has been shown to reduce the physiological markers of stress, such as heart rate and cortisol levels, while boosting dopamine levels in the brain. The calming effect of TM enhances

, promoting better mental health and emotional balance.

Neurofeedback and Biofeedback for Optimizing Dopamine Activity

While mindfulness and meditation directly influence dopamine regulation, **neurofeedback** and **biofeedback** represent cutting-edge approaches to actively monitor and modulate dopamine activity in real-time. These technologies offer a more **personalized** and **precise** way of balancing dopamine levels to optimize mental and emotional performance.

Neurofeedback

- Neurofeedback involves the use of **electroencephalography (EEG)** to monitor brainwave patterns. By receiving real-time feedback on brain activity, individuals can learn to regulate their brainwaves, optimizing dopamine pathways involved in focus, relaxation, and emotional stability.
- Neurofeedback has been shown to increase **dopamine receptor sensitivity**, particularly in areas linked to **attention**, **executive functions**, and **motivation**. Regular training can result in improved mental clarity, emotional control, and cognitive flexibility.

Biofeedback

- Biofeedback, which involves monitoring physiological responses such as heart rate, skin temperature, and muscle tension, can also help individuals regulate their **dopamine** levels. By learning to control these physiological markers, biofeedback allows individuals to cultivate a state of **relaxed alertness**, promoting healthy dopamine function.
- Studies suggest that biofeedback can help regulate **dopamine-driven behaviors**, such as impulse control and emotional responses, leading to improved **mental resilience** and **mood regulation**.

Practical Tips for Integrating Mindfulness and Meditation into Daily Life

1. **Start Small**: If you're new to mindfulness or meditation, start with short sessions—just 5-10 minutes per day. Gradually increase the duration as you become more comfortable. Even brief moments of mindfulness can have significant effects on dopamine regulation.
2. **Consistency Is Key**: Like any practice, mindfulness and meditation require consistency to be effective. Make these practices part of your daily routine to experience long-term benefits for dopamine balance and overall mental health.
3. **Use Guided Meditations**: If you're unsure where to begin, try using apps or online resources that offer guided meditation. These can help keep you focused and provide structure, making the process easier and more accessible.
4. **Incorporate Mindfulness into Everyday Activities**: You don't need to meditate for hours to benefit from mindfulness. Try practicing mindfulness during daily activities, such as eating, walking, or washing dishes. The key is to bring **full awareness** to the present moment.
5. **Monitor Your Emotional States**: Pay attention to your emotional responses throughout the day. Notice when you're feeling stressed or overwhelmed, and take a few minutes to practice mindfulness to bring yourself back to balance.

Conclusion

Mindfulness and **meditation** are not just tools for stress reduction—they are powerful practices for **balancing dopamine levels** and optimizing cognitive and emotional function. By enhancing **dopamine receptor activity**—particularly **DRD1** and **DRD2**—these mental practices foster greater focus, emotional resilience, and mental clarity. Whether through **mindfulness meditation, Loving-Kindness Meditation**, or **neurofeedback**, individuals can harness the power of dopamine regulation to lead more fulfilling, balanced lives. As we delve deeper into these practices, we unlock the potential to live not just mindfully but with greater cognitive and emotional mastery.

Chapter 17: Social Connection and Dopamine – Building Healthy Relationships

Human beings are inherently social creatures. From the earliest stages of life, our ability to connect with others shapes our mental and emotional health. The concept of **dopamine** is crucial in understanding the physiological processes underlying these relationships, as it is directly tied to our experiences of reward, pleasure, and motivation. In this chapter, we explore the intricate ways in which social interactions influence dopamine release, the roles of **DRD1** and **DRD2** in forming and maintaining relationships, and how fostering positive social environments can support dopamine balance and overall well-being.

The Science Behind Social Connection and Dopamine Release

Dopamine plays a fundamental role in how we **experience pleasure** from social interactions. Whether through **touch, conversation, laughter,** or **shared activities**, social connection triggers the release of dopamine, which reinforces the rewarding nature of these experiences. This release is crucial not only for forming bonds but also for **maintaining motivation** and emotional stability.

Dopamine and the Reward System

- Social interactions activate the brain's **reward system**, particularly the **striatum**, which is involved in the processing of rewards and reinforcement. When we engage in positive social experiences, dopamine is released, reinforcing the desire to continue interacting and deepening the bond with others. These experiences include time spent with loved ones, positive feedback, shared achievements, or even simply being present in a supportive social environment.
- This activation of dopamine reinforces behaviors that increase social bonding and emotional connection, helping to build and sustain relationships over time.

The Role of DRD1 and DRD2 in Social Bonding

- Both **DRD1** and **DRD2** receptors play significant roles in regulating the brain's reward and motivation systems. **DRD1**, in particular, is associated with **motivation, goal-setting**, and **cognitive flexibility**. Positive social interactions, such as achieving shared goals with others or receiving support, stimulate **DRD1** activation, reinforcing the motivation to engage in more social activities.
- On the other hand, **DRD2** is linked to the experience of **pleasure** and **reward**. It governs the reinforcement of pleasurable experiences, including emotional rewards from **attachment** and **affiliation**. Higher sensitivity of **DRD2** receptors has been associated with better emotional regulation and greater satisfaction in social relationships.

How Social Support Enhances Dopamine Balance

One of the most significant ways that social connections impact dopamine balance is through **social support**. Having a **strong social network**—whether through family, friends, or peers—provides a buffer against stress and promotes mental and emotional resilience.

Social Support and Stress Reduction

- Chronic stress has a detrimental impact on dopamine signaling, often leading to a depletion of dopamine and dysfunction of the dopamine system. However, positive social interactions can mitigate this stress, reducing the impact of cortisol (the stress hormone) and stimulating dopamine release. This allows for better emotional regulation and an enhanced sense of well-being.
- Studies have shown that people with strong social connections tend to experience **lower levels of stress** and **greater overall happiness**, partly due to the protective effects of dopamine in response to social support.

Building Trust and Cooperation

- **Trust** is fundamental in any meaningful social relationship. When we trust others, dopamine is released as part of the brain's reward system. This promotes a sense of **security** and **connection**, which is vital for maintaining healthy relationships.
- Cooperative behavior in social interactions is similarly linked to dopamine. When individuals collaborate to achieve mutual goals, their brain's reward centers are activated, reinforcing cooperation and mutual support. This creates a positive feedback loop that encourages further positive interactions and bonding.

The Impact of Positive Social Environments on Dopamine Regulation

Creating and maintaining a **dopamine-friendly social environment** is crucial for promoting mental and emotional health. The social environment not only influences dopamine levels but also shapes how individuals interact with their world.

Fostering Positive Social Connections

- Positive social environments are characterized by **mutual respect, support**, and **shared positivity**. These types of interactions trigger dopamine release and contribute to a sense of belonging. Whether through family gatherings, close friendships, or community activities, regular participation in these environments helps maintain balanced dopamine levels and strengthens **social bonds**.
- **Social rituals**—such as shared meals, celebrations, or collaborative activities—can reinforce feelings of belonging, which are crucial for maintaining emotional balance and fostering healthy relationships. These rituals not only promote social cohesion but also stimulate **DRD2** receptor activity, reinforcing positive emotional states.

Avoiding Toxic Social Environments

- Just as positive social connections can enhance dopamine function, toxic or negative environments can have the opposite effect. Social **stressors** such as **conflict, betrayal**, or **isolation** can increase stress hormones like cortisol, which inhibit dopamine production and receptor sensitivity. In turn, this can lead to **dopamine dysregulation**, contributing to feelings of anxiety, depression, and emotional imbalance.
- It's essential to identify and distance oneself from negative influences and instead cultivate relationships with individuals who prioritize **empathy**, **understanding**, and **mutual respect**.

Building Healthy Relationships: Practical Tips for Optimizing Dopamine Balance

To fully leverage the benefits of social connection on dopamine balance, individuals can actively work toward building and nurturing positive relationships. Here are a few practical strategies:

Prioritize Quality over Quantity

Rather than focusing on the number of people in your social circle, prioritize

. Cultivating deep, meaningful relationships with a few trusted individuals is far more beneficial than having a large but shallow network. These deep connections are more likely to stimulate dopamine release and foster lasting emotional satisfaction.

Engage in Shared Activities

Participating in activities that involve cooperation or shared goals—such as sports, creative projects, or volunteer work—can enhance dopamine release and strengthen social bonds. These experiences provide opportunities for both

-related motivation and

-related pleasure.

Practice Active Listening and Empathy

Active listening, where you focus on understanding the other person's perspective without judgment, helps build trust and emotional connection. This strengthens relationships and triggers dopamine release, enhancing both the giver's and receiver's feelings of reward.

Express Appreciation and Gratitude

Regularly expressing gratitude and appreciation in relationships fosters

. Simple acts of kindness, such as thanking others or acknowledging their efforts, can trigger dopamine release, reinforcing the bond between individuals.

Seek Emotional Balance through Social Feedback

Social feedback, such as receiving encouragement or praise from others, can help regulate dopamine levels. Constructive feedback fosters a sense of accomplishment and motivates individuals to continue engaging in positive behaviors. Ensure that your social environment is supportive and provides the feedback you need to grow emotionally and cognitively.

Conclusion

Social connection is more than just a means of passing time—it's a fundamental driver of **dopamine release, emotional stability**, and **cognitive function**. Through **positive interactions, supportive environments**, and **trust-building activities**, we can optimize dopamine balance, enhance motivation, and deepen our relationships. By consciously nurturing our social circles and maintaining healthy, supportive interactions, we can unlock the power of dopamine to foster a sense of belonging, joy, and resilience. Whether through small acts of kindness, shared experiences, or consistent emotional support, the social connections we cultivate can directly impact our dopamine regulation, contributing to a healthier, more balanced life.

Chapter 18: Dopamine and Creativity – Unlocking Your Creative Potential

Creativity is a powerful force that drives innovation, problem-solving, and personal expression. At its core, creativity is not just about artistic endeavors—it is a fundamental aspect of human cognition, influencing everything from scientific breakthroughs to everyday decision-making. But what makes us creative? How can we tap into this potential and unlock the full capacity of our creative minds?

The answer may lie in the **dopamine system**, specifically the **DRD1** and **DRD2** receptors. Dopamine, the brain's primary reward neurotransmitter, has a profound impact on creative thinking and problem-solving. By understanding how **dopamine receptors** influence creativity, we can learn how to optimize our mental processes, overcome cognitive blocks, and foster innovative ideas.

In this chapter, we will explore the relationship between **dopamine** and **creativity**, focusing on the roles of **DRD1** and **DRD2** in **divergent thinking, cognitive flexibility**, and **the creative process**. We will also discuss practical strategies for enhancing creativity through dopamine optimization.

The Science of Dopamine and Creativity

Dopamine has long been linked to the brain's **reward system** and motivation pathways. However, it also plays a crucial role in **cognitive flexibility** and the ability to **think outside the box**—key attributes of creativity. Understanding the way dopamine functions in the creative process begins with examining how dopamine receptors **DRD1** and **DRD2** support various cognitive mechanisms.

DRD1 and Divergent Thinking

- **Divergent thinking** is the ability to generate multiple solutions to an open-ended problem—one of the hallmarks of creativity. DRD1 receptors are closely associated with **cognitive flexibility**, which allows the brain to switch between different modes of thinking and consider alternative solutions.
- Activation of **DRD1** receptors enhances the brain's capacity for **exploration**—the ability to generate a variety of possibilities without being constrained by past experiences or existing knowledge. This makes **DRD1** central to the **creative process**, especially in early stages when new, untested ideas emerge.

DRD2 and Reward-Driven Creativity

- **DRD2** receptors are deeply involved in the **experience of pleasure** and reward, which are key motivators for creative engagement. Creative individuals are often driven by a sense of reward—whether intrinsic (the joy of creating) or extrinsic (recognition and success). The activation of **DRD2** encourages continued engagement in creative tasks by reinforcing pleasurable experiences associated with the creative process.
- Studies have shown that **DRD2 sensitivity** is linked to the brain's ability to sustain focus and **persistence** in creative work, even in the face of challenges or setbacks. This receptor also plays a role in the **reward feedback loop**, which is activated when creative efforts lead to successful outcomes.

How Dopamine Influences Problem-Solving and Innovation

Creativity is not just about generating ideas; it is also about solving problems and adapting to new challenges. Dopamine, through its interaction with **DRD1** and **DRD2**, facilitates this complex cognitive process in the following ways:

Enhanced Problem-Solving Abilities

- Dopamine supports cognitive processes that are vital for **problem-solving**, including **working memory, attention,** and **mental flexibility**. DRD1's involvement in cognitive flexibility allows individuals to approach problems from different angles, while **DRD2** contributes by reinforcing behaviors that lead to **successful solutions**.
- A well-balanced dopamine system is crucial for creative problem-solving because it enables individuals to move fluidly between divergent thinking (coming up with multiple ideas) and convergent thinking (narrowing those ideas down to the best solution).

The Creative Cycle – From Insight to Innovation

- Creativity often follows a cyclical pattern that involves **insight, trial and error**, and **adjustment**. Dopamine is involved in each phase of this cycle, facilitating insight by supporting the brain's ability to make novel connections, as well as maintaining motivation through the process of experimentation.
- Activation of **DRD2** receptors ensures that the brain receives positive reinforcement for creative attempts, fueling the individual's motivation to push through failure and pursue novel solutions until they achieve innovation.

Optimizing Dopamine for Creative Flow States

A key to unlocking creativity is achieving a state of **flow**—the optimal psychological state in which a person is fully immersed in an activity, experiencing deep focus, enjoyment, and intrinsic motivation. Flow is associated with heightened dopamine release and optimal brain function, and understanding how to enhance dopamine signaling can help individuals enter and sustain flow states more easily.

The Role of Dopamine in Flow

Flow is facilitated by a delicate balance of dopamine, where the brain is simultaneously focused and relaxed, maintaining a high level of performance without becoming overwhelmed or fatigued. In this state, both and

play crucial roles:

- **DRD1** is linked to the brain's **reward-seeking behavior**, motivating individuals to stay engaged in the creative task.
- **DRD2** reinforces the feeling of satisfaction as the individual progresses, allowing the mind to remain in the flow state longer.

Triggering Flow States

Certain conditions can help trigger flow states and optimize dopamine release. These include:

- **Challenging yet achievable goals**: Tasks that are just within one's skill level but still require effort trigger dopamine release.
- **Clear feedback**: Immediate feedback on performance helps maintain dopamine levels, reinforcing the creative process.
- **Immersive environments**: Environments that minimize distractions and allow full focus on the task enhance dopamine flow and support sustained creativity.

Practical Strategies for Enhancing Creativity through Dopamine Optimization

While dopamine plays a central role in creativity, there are actionable strategies that individuals can adopt to **optimize dopamine function** and enhance creative output:

Engage in Regular Mental Challenges

Creativity thrives on

. Regularly challenging your mind with puzzles, brainstorming sessions, or novel tasks can increase dopamine release and enhance cognitive flexibility. The key is to maintain a balance between difficulty and ability—too easy a task won't engage the brain enough, while too difficult a task might overwhelm it.

Mindful Practices for Creativity

- Practices like **meditation**, **deep breathing**, and **mindfulness** can help manage dopamine levels by reducing stress and promoting a balanced state of alertness and relaxation. Mindfulness also supports divergent thinking by allowing the brain to remain open to new ideas without judgment.
- **Visualization** techniques, often used by artists and athletes, can also stimulate dopamine pathways, preparing the mind for creative activity by reinforcing the brain's reward system.

Physical Exercise for Dopamine Boost

- Physical activity, especially **aerobic exercises** like running, cycling, or swimming, has been shown to boost dopamine levels and receptor sensitivity. Exercise not only increases dopamine production but also promotes the growth of new brain cells in areas associated with creativity and problem-solving.
- Taking regular breaks to engage in physical activity can also help clear mental blockages and renew creative energy.

Incorporate Novelty and Exploration

- Seeking new experiences and stepping outside your comfort zone activates the dopamine system and encourages **new neural connections**. Try exploring new hobbies, taking up new creative practices, or immersing yourself in unfamiliar environments to foster creativity.
- The brain thrives on **novelty**—whether it's new challenges, environments, or ideas. Engage in activities that push your boundaries and stimulate your curiosity to encourage fresh ideas and creative breakthroughs.

Conclusion

Creativity is a complex and multifaceted process that is deeply intertwined with the brain's dopamine system. By understanding how **DRD1** and **DRD2** receptors influence cognitive flexibility, problem-solving, and reward-driven behavior, we can take intentional steps to optimize our brain's creative potential. Through practical strategies such as engaging in mental challenges, practicing mindfulness, exercising regularly, and embracing novelty, we can cultivate a creative mindset that fosters innovation and problem-solving in all aspects of life.

Ultimately, creativity is not just a gift—it's a skill that can be nurtured and enhanced through an understanding of the brain's biochemical processes. By unlocking the power of dopamine receptors, we can empower ourselves to reach new heights of creative potential, transforming both our personal and professional lives.

Chapter 19: Advanced Techniques – Targeting Dopamine Pathways for Personal Growth

As our understanding of dopamine and its receptors—**DRD1** and **DRD2**—deepens, it opens up a world of possibilities for enhancing mental and emotional well-being. While much of this book has focused on the fundamentals of dopamine regulation, this chapter will take a deeper dive into **advanced techniques** that target dopamine pathways for **personal growth**. From cognitive strategies to biohacking and nootropics, we will explore cutting-edge methods to optimize dopamine signaling and enhance performance, creativity, and emotional balance.

By leveraging these advanced techniques, individuals can fine-tune their dopamine systems, promoting sustained motivation, emotional regulation, and cognitive clarity. The potential to influence dopamine pathways offers exciting opportunities for anyone looking to push the boundaries of their personal development.

1. Cognitive Strategies to Manipulate Dopamine Signaling

To unlock the full potential of dopamine pathways, it is essential to engage in cognitive strategies that optimize receptor function. By understanding how **dopamine regulates behavior**, we can adopt mental practices that enhance focus, motivation, and creativity.

Cognitive Behavioral Strategies (CBT) for Dopamine Regulation:

- **Goal Setting**: Dopamine is a key player in motivation, and setting achievable goals can stimulate its release. Break large goals into smaller, more manageable steps to keep the brain engaged and motivated. Celebrating small wins will trigger dopamine release, reinforcing progress and creating a positive feedback loop.
- **Visualization**: The act of mentally rehearsing success—whether it's finishing a project, achieving a goal, or experiencing a rewarding outcome—activates dopamine pathways associated with motivation and goal-directed behavior. Visualizing the rewards you'll experience once a task is complete can boost dopamine levels and enhance focus.
- **Positive Reinforcement**: Cultivate an environment where positive reinforcement is frequent. Reward yourself for meeting milestones or achieving positive behavioral changes. This not only strengthens dopamine signaling but also fosters a mindset focused on progress and growth.

Mindset Shifts:

Developing a

is vital for long-term cognitive and emotional optimization. This mindset, which embraces challenges and sees failure as an opportunity to learn, activates dopamine release in response to perseverance and effort. A

, by contrast, can stunt motivation, leading to reduced dopamine activity.

Self-Talk and Dopamine:

Engaging in

—reminding yourself of your capabilities and reinforcing your potential—can increase dopamine signaling. On the flip side, negative self-talk can inhibit dopamine release, limiting motivation and creativity. By training yourself to speak positively and optimistically, you can create a mental environment conducive to dopamine activation.

2. Biohacking Dopamine for Peak Performance

Biohacking has gained significant popularity as a method to optimize physical and mental performance. By targeting the **dopamine system** directly, individuals can accelerate cognitive growth, emotional stability, and creative potential.

Nootropics and Supplements:

Nootropics are substances known to enhance cognitive function, and many of them influence dopamine pathways. Some of the most popular

include:

- **L-Tyrosine**: A precursor to dopamine, L-Tyrosine supports dopamine synthesis and can improve focus and cognitive performance, particularly in stressful situations.
- **Rhodiola Rosea**: This adaptogen has been shown to enhance dopamine activity and help the brain adapt to stress, improving mental clarity and resilience.
- **Mucuna Pruriens**: This plant extract contains **L-Dopa**, a direct precursor to dopamine. It can elevate dopamine levels, improve mood, and enhance cognitive function, making it particularly useful for individuals with low dopamine tone or those seeking cognitive enhancement.
- **Curcumin**: Known for its anti-inflammatory properties, curcumin has been shown to support the release of dopamine and enhance memory function. Regular supplementation may protect against age-related cognitive decline by preserving dopamine receptor activity.

Exercise and Dopamine Regulation:

- Physical activity has profound effects on dopamine signaling. **Aerobic exercises** like running, swimming, or cycling increase dopamine receptor density, improving mood and cognitive function. Studies have shown that regular exercise boosts dopamine release, improving focus and motivation.
- Incorporating **high-intensity interval training (HIIT)** or **strength training** into your routine can provide additional dopamine benefits. These intense bursts of activity not only stimulate dopamine but also promote neuroplasticity, encouraging the formation of new neural connections in areas related to cognitive flexibility and creativity.

Sleep Optimization:

- Sleep is one of the most powerful tools for optimizing dopamine function. During sleep, especially during **REM sleep**, the brain undergoes critical processes of synaptic pruning and dopamine receptor repair. **Sleep deprivation**, on the other hand, can significantly reduce dopamine receptor sensitivity, leading to decreased motivation, mood disturbances, and cognitive fog.
- Prioritize **quality sleep** by maintaining a regular sleep schedule, avoiding stimulants like caffeine in the evening, and engaging in relaxing pre-sleep routines to optimize dopamine receptor function.

3. Neurofeedback and Brainwave Optimization

Advances in neurotechnology have given rise to tools like **neurofeedback**, which can be used to train the brain to function optimally. Neurofeedback allows individuals to monitor their brainwave activity and make adjustments in real time to enhance dopamine regulation.

How Neurofeedback Works:

- Neurofeedback involves placing sensors on the scalp to measure brainwave activity. The feedback provided to the individual allows them to adjust their mental state by consciously shifting brainwave patterns. By optimizing the **theta**, **alpha**, and **beta** brainwaves, individuals can increase **dopamine release** in regions of the brain involved in motivation, learning, and creativity.
- Studies have shown that **neurofeedback training** can improve executive function, emotional regulation, and cognitive flexibility by directly influencing the dopamine system.

Brainwave Entrainment:

Brainwave entrainment

alpha waves

beta waves

4. Combining Lifestyle Practices for Optimal Dopamine Regulation

Beyond biohacking and advanced techniques, integrating **holistic lifestyle practices** is essential for optimizing dopamine pathways. Small adjustments in daily life can make a big difference in overall dopamine function and well-being.

Mindfulness and Meditation:

- Meditation practices, especially **mindfulness meditation**, have been shown to modulate dopamine receptors. By training the mind to focus on the present moment, individuals can increase dopamine sensitivity, promoting emotional regulation and cognitive clarity.
- **Loving-kindness meditation**, which fosters positive emotions and compassion, has also been linked to increased dopamine release, improving both mood and social connection.

Diet and Nutrition:

- A **dopamine-optimized diet** rich in **tyrosine**, the precursor to dopamine, supports receptor function. Foods like lean meats, eggs, dairy, and soy products provide essential amino acids that promote dopamine synthesis.
- Healthy fats, such as those found in avocados, nuts, and fish, support brain health and dopamine receptor function. Additionally, consuming **antioxidants** from fruits and vegetables can protect dopamine receptors from oxidative damage, supporting long-term mental and emotional well-being.

Stress Management:

Chronic stress can

, leading to fatigue, burnout, and decreased motivation. Incorporating

, such as yoga, deep breathing, or spending time in nature, can help restore balance to the dopamine system and improve overall mental health.

Conclusion

Mastering dopamine pathways for personal growth is not just about understanding the science of dopamine receptors—it's about applying that knowledge in practical, advanced ways to optimize cognitive, emotional, and behavioral outcomes. By utilizing **cognitive strategies**, **biohacking techniques**, and **neurofeedback**, individuals can take control of their dopamine systems, leading to enhanced creativity, emotional stability, and peak performance.

As we continue to explore the potential of dopamine modulation, the ability to optimize our brain's reward system opens new frontiers in personal development. These advanced techniques, when applied thoughtfully and responsibly, offer an exciting path toward mastering the art of human potential and unlocking the power of dopamine for lasting growth and success.

Chapter 20: The Ethics of Dopamine Manipulation – Balancing Enhancement with Responsibility

As we explore the power of dopamine and its receptors—particularly **DRD1** and **DRD2**—to optimize mental and emotional well-being, we must consider the ethical implications of manipulating these pathways. Dopamine plays a critical role in shaping behavior, motivation, mood, and cognitive function, and the ability to regulate or enhance dopamine signaling offers vast potential for personal growth. However, this power comes with ethical responsibilities.

In this chapter, we will examine the ethical concerns surrounding dopamine manipulation, including the risks of over-activation, dependency, and the potential for misuse. We will explore how to balance the desire for enhanced performance and emotional balance with the need for responsible, mindful application of dopamine-boosting techniques. Ultimately, this chapter aims to provide guidance on how to responsibly harness the power of dopamine for individual and societal benefit.

1. The Power and Potential of Dopamine Enhancement

Dopamine's role as a **reward molecule** is undeniable. It drives motivation, reinforces pleasurable experiences, and facilitates learning. As such, there is immense appeal in manipulating dopamine levels to achieve higher cognitive performance, emotional regulation, and productivity. From the use of nootropics to biohacking techniques that influence dopamine receptor sensitivity, the potential for enhancing human performance is vast.

However, with the power to boost dopamine comes the responsibility to use these tools ethically. It's crucial to consider the broader impacts that **dopamine manipulation** may have, not only on individual users but on society at large.

Personal Growth and Enhancement:

- Many individuals seek to optimize their mental and emotional states using dopamine-enhancing techniques. Whether it's through cognitive strategies, physical activity, diet, or pharmacological interventions, the goal is often to achieve peak performance or emotional balance. This is especially true for individuals who may suffer from **dopamine dysregulation** due to stress, mental health issues, or age-related decline.
- While the benefits of dopamine manipulation are clear, the question remains: **How much is too much?** Where do we draw the line between beneficial enhancement and potential harm? For example, while nootropics like L-Tyrosine and Mucuna Pruriens can enhance cognitive function, overuse or improper dosing can lead to overstimulation of dopamine receptors, which may result in negative effects such as anxiety, impulsivity, and even addiction.

2. The Risks of Over-Activation

One of the primary concerns with manipulating dopamine receptors is the risk of **over-activation**. Dopamine is a double-edged sword—while it can foster motivation and focus, excessive dopamine release or overstimulation of **DRD1** and **DRD2** receptors can lead to detrimental effects on mental and physical health.

Addiction and Dependency:

- Dopamine is closely linked to **reward-seeking behavior**. The brain's reward system is designed to encourage behaviors that promote survival and reproduction, such as eating, socializing, and procreating. However, substances or behaviors that artificially stimulate dopamine—such as drugs, gambling, or even social media use—can hijack this system, leading to **dopamine addiction**.
- **Substance use disorders** (e.g., cocaine, methamphetamine) can cause profound shifts in the dopamine system, leading to an impaired ability to regulate dopamine release and receptor sensitivity. Individuals who manipulate dopamine pathways using nootropics, stimulants, or other substances may inadvertently alter their brain's natural dopamine balance, risking dependence on external stimuli to maintain motivation or mood stability.

Psychological Implications:

- Over-activation of dopamine receptors may lead to **impulsivity, irrational decision-making**, and **emotional instability**. When dopamine levels are chronically elevated, it can result in an inability to regulate emotions, creating a cycle of heightened arousal and emotional dysregulation.
- High levels of dopamine are also linked to the **reward prediction error**—a phenomenon in which expectations of pleasure or reward consistently exceed actual experiences. This creates a sense of **dissatisfaction** or **anxiety**, as the brain constantly chases rewards that may never fully materialize.

Neurotoxicity:

Chronic overstimulation of dopamine receptors can lead to

, damaging the neural circuits that regulate mood, learning, and memory. Over time, this can impair cognitive function and emotional regulation, making it harder for individuals to achieve the very goals they were initially trying to enhance.

3. The Ethics of Cognitive Enhancement

With the rise of biohacking and neuroenhancement technologies, the ethical concerns surrounding **cognitive enhancement** have become increasingly significant. As we have seen, manipulating dopamine levels can offer a competitive edge, whether in academics, business, or athletic performance. However, the question of whether such enhancements are **fair**, **just**, or even **morally acceptable** remains a topic of debate.

Equity and Access:

- One major ethical concern is **accessibility**. If certain methods of dopamine manipulation—such as nootropics or cognitive-enhancing drugs—become widely used, they may create a divide between those who can afford these tools and those who cannot. This could exacerbate existing inequalities, particularly in education, career advancement, and social mobility.
- Access to biohacking tools may not be equally available to all socioeconomic groups, potentially creating a **two-tiered society** where those with resources have an enhanced cognitive or emotional advantage over others.

Informed Consent:

Another concern is the issue of

. As dopamine manipulation techniques become more popular, individuals may not fully understand the long-term consequences of their actions. Misuse of nootropics or the overuse of stimulant medications can lead to

, including addiction, cognitive decline, and mental health issues. Ethical biohacking demands that individuals are fully informed about the risks and benefits of any dopamine-altering techniques they choose to engage in.

Social and Psychological Pressure:

As cognitive enhancement becomes more normalized, there is the potential for

 to participate. In environments like work or school, individuals may feel compelled to enhance their cognitive function artificially to keep up with peers or meet high expectations. This pressure can erode the concept of natural talent and effort, leading to a

 that undermines well-being.

4. Responsible Use of Dopamine-Enhancing Techniques

To navigate the ethical landscape of dopamine manipulation, it is crucial to adopt a framework that promotes **responsible use** of dopamine-enhancing techniques. Below are several guidelines for individuals looking to optimize their dopamine systems while minimizing potential harm:

Balance Over Enhancement:

While there is value in enhancing dopamine function to improve motivation, cognitive performance, and emotional regulation, the goal should be

 rather than

. The pursuit of long-term cognitive and emotional health should take precedence over short-term gains in performance.

Mindful Practices:

Mindfulness, self-awareness, and regular self-reflection can help individuals become more attuned to how their dopamine systems are functioning. By being mindful of the effects of dopamine manipulation—both positive and negative—individuals can make more informed decisions about how and when to engage in enhancement practices.

Gradual and Temporary Approaches:

Dopamine-enhancing techniques should be approached gradually, with careful consideration of their impact on the brain and overall well-being. Temporary boosts in dopamine levels—such as those achieved through exercise, meditation, or nootropic use—should be used strategically rather than continually or excessively.

Restoring Balance:

Ultimately, the ability to balance dopamine—recognizing when it is overactive and when it is underactive—is crucial.

, such as taking breaks from stimulation, engaging in social connection, and practicing relaxation techniques, can help restore natural dopamine balance and prevent burnout or overstimulation.

5. The Future of Ethical Dopamine Manipulation

As dopamine research and biohacking technologies continue to evolve, it is essential for society to engage in a broader conversation about the ethics of cognitive and emotional enhancement. The ability to manipulate dopamine pathways offers tremendous opportunities for improving mental health, performance, and well-being, but these opportunities must be approached with caution and responsibility.

Regulation and Oversight:

As new technologies and substances emerge, there will be a growing need for

 to ensure that they are used ethically and safely. This includes developing guidelines for the responsible use of nootropics, stimulants, and other dopamine-altering agents. By establishing clear regulations, society can help protect individuals from harm while allowing for innovation and progress in dopamine research.

Education and Awareness:

Educating the public about the

 of dopamine-enhancing techniques is essential. Awareness campaigns, academic research, and public discourse can help individuals make informed choices and avoid the pitfalls of over-manipulating dopamine pathways.

A Balanced Future:

The future of dopamine manipulation lies in achieving a balance between enhancement and well-being. As our understanding of dopamine receptors like

and

deepens, we can expect new methods of optimizing mental health and performance. However, this must be done in a way that respects the natural complexities of the human brain, ensuring that enhancements contribute to, rather than detract from, the well-being of individuals and society as a whole.

Conclusion

The ethical concerns surrounding dopamine manipulation are complex and multifaceted, but they are an integral part of our journey to understanding the brain's reward system. While the potential to enhance cognitive and emotional states is compelling, it is important to approach dopamine manipulation with **responsibility** and **awareness**. By considering the risks, ensuring equitable access, and striving for a balanced approach, we can harness the power of dopamine to unlock human potential in a way that is both **ethical** and **sustainable**.

Chapter 21: The Future of Dopamine Research – Cutting-Edge Discoveries and Innovations

The science of dopamine and its receptors, particularly **DRD1** and **DRD2**, has already transformed our understanding of brain function, mental health, and behavior. However, we are just beginning to scratch the surface of the vast potential that lies in the manipulation and understanding of these receptors. In this chapter, we will explore the latest trends and cutting-edge discoveries in dopamine research, as well as the groundbreaking innovations that promise to shape the future of neuroscience and mental health treatment.

As we venture into the future, the manipulation of dopamine systems holds both exciting potential and significant challenges. This chapter will highlight the advancements in **dopamine receptor modulation**, the development of new **pharmacological treatments**, and the promise of **personalized medicine**. We'll also examine how these innovations could lead to improved therapies for addiction, neurodegenerative diseases, mood disorders, and cognitive impairments, as well as the ethical considerations that accompany them.

1. Emerging Trends in Dopamine Research

Dopamine research is currently undergoing a transformation, driven by new technologies, genetic insights, and experimental methodologies. These trends point to a future where our understanding of dopamine signaling and its impact on behavior could lead to personalized, more effective therapies.

a. Genetic and Epigenetic Insights:

- One of the most exciting developments in dopamine research is the growing understanding of the **genetic and epigenetic factors** that influence dopamine receptors. Scientists are discovering how genetic variations in the **DRD1** and **DRD2** genes may affect receptor function and influence susceptibility to mental health disorders such as depression, schizophrenia, and addiction.
- **Epigenetics**, or the study of how environmental factors influence gene expression, is also revealing how lifestyle choices, stress, and diet can impact dopamine receptor function. These insights are paving the way for more targeted interventions that consider not only genetic predispositions but also the role of external factors in shaping dopamine systems.

b. Advanced Imaging Technologies:

- Cutting-edge brain imaging techniques, such as **positron emission tomography (PET)** and **functional magnetic resonance imaging (fMRI)**, are providing unprecedented views of dopamine activity in the brain. These technologies allow scientists to observe how dopamine receptors function in real-time, helping to identify abnormalities in brain regions associated with reward, motivation, and mood regulation.
- Future advancements in imaging will likely lead to even more precise mapping of dopamine pathways, enabling personalized interventions for a wide range of psychiatric and neurological conditions. This could open the door to individualized treatment plans tailored to a person's unique dopamine profile.

c. Neural Circuitry and Dopamine Modulation:

- Researchers are increasingly focusing on the **neural circuits** that involve dopamine signaling. The brain is not just a collection of isolated regions; it is a complex network where dopamine interacts with other neurotransmitters like serotonin, glutamate, and GABA. Understanding how dopamine fits into this larger framework is crucial for developing more effective treatments for conditions like addiction, depression, and Parkinson's disease.
- **Optogenetics**, a technology that uses light to control cells within living tissue, is enabling scientists to manipulate dopamine neurons in specific brain regions with incredible precision. This could eventually lead to novel treatments for conditions that involve disrupted dopamine signaling, such as addiction, bipolar disorder, or even the cognitive decline associated with aging.

2. Breakthroughs in Dopamine-Related Therapies

The future of dopamine research holds the promise of groundbreaking therapies that could revolutionize the treatment of various neuropsychiatric and neurodegenerative disorders. Researchers are exploring new avenues for treating conditions linked to dopamine dysregulation, including addiction, schizophrenia, depression, and Parkinson's disease. These therapies may target both **DRD1** and **DRD2**, as well as the broader dopamine system, to rebalance dopamine function and improve patients' quality of life.

a. Personalized Medicine:

- One of the most promising developments in dopamine-related therapies is **personalized medicine**—treatments that are tailored to the individual based on their unique genetic makeup, lifestyle, and environmental factors. For example, researchers are developing genetic tests that can identify specific mutations or variations in dopamine receptor genes, such as DRD1 and DRD2, that may influence how a person responds to treatment.
- Personalized medicine also extends to **pharmacogenomics**, the study of how genetic variations affect an individual's response to drugs. This approach could lead to more effective treatments for conditions like addiction or depression, as clinicians could choose the most appropriate drugs based on a patient's specific genetic profile.

b. Targeted Pharmacological Interventions:

- Traditional treatments for dopamine-related disorders, such as **antipsychotics** or **stimulants**, are often blunt instruments that can have broad and sometimes harmful side effects. However, researchers are working on **targeted therapies** that specifically modulate DRD1 and DRD2 receptor activity without affecting other parts of the dopamine system.
- **Selective dopamine receptor agonists** and **antagonists** are being developed to precisely modulate dopamine signaling. For example, **DRD1 agonists** could potentially be used to enhance motivation and cognitive flexibility in individuals with conditions like depression or schizophrenia. On the other hand, **DRD2 antagonists** could be used to treat addiction by decreasing the rewarding effects of substances like cocaine or nicotine.

c. Non-Pharmacological Approaches:

- In addition to pharmaceutical interventions, there is increasing interest in non-pharmacological approaches that can influence dopamine receptor function. **Neurostimulation techniques** such as **transcranial magnetic stimulation (TMS)** and **deep brain stimulation (DBS)** are being explored for their potential to modulate dopamine activity and treat conditions like depression, Parkinson's disease, and addiction.
- **Cognitive-behavioral therapies (CBT)** and **mindfulness-based interventions** are also being studied for their ability to rebalance dopamine signaling, particularly in the context of emotional regulation. These therapies help individuals rewire neural circuits that may have become dysfunctional due to chronic stress or maladaptive behavior patterns.

3. The Role of Artificial Intelligence and Big Data

The future of dopamine research is also being shaped by **artificial intelligence (AI)** and **big data**. These technologies are providing researchers with the tools to process vast amounts of data and identify patterns in dopamine function that were previously difficult to detect.

a. AI-Driven Drug Discovery:

- AI is revolutionizing the field of drug discovery by enabling researchers to rapidly analyze large datasets of molecular information. In the context of dopamine research, AI algorithms can sift through massive databases of potential drug candidates to identify compounds that might modulate DRD1 and DRD2 activity in specific ways.
- AI is also helping to predict how new drugs might interact with dopamine receptors, accelerating the development of targeted therapies for conditions like addiction, Parkinson's disease, and schizophrenia.

b. Data-Driven Neuroscience:

- Big data is playing an increasingly important role in neuroscience by enabling researchers to analyze massive datasets of brain activity, genetic data, and behavioral patterns. By collecting and analyzing this data, researchers can uncover new insights into how dopamine functions at the level of individual neurons, neural circuits, and brain regions.
- This data-driven approach also allows scientists to create more accurate models of dopamine signaling, leading to better predictions about how specific interventions—whether pharmacological or behavioral—will impact dopamine function in the brain.

4. Ethical Considerations for the Future

While the future of dopamine research holds immense potential, it also presents significant ethical challenges. As we gain the ability to more precisely manipulate dopamine receptors, we must carefully consider the implications of such interventions.

a. Equity and Access:

As new dopamine-related therapies and technologies emerge, it will be essential to ensure that these advancements are accessible to all individuals, regardless of socioeconomic status. If these therapies are limited to certain populations, there is the potential to widen health disparities and create a divide between those who can afford cutting-edge treatments and those who cannot.

b. Safety and Long-Term Effects:

The safety of dopamine manipulation technologies must be rigorously tested, particularly as new pharmacological agents, neurostimulation techniques, and genetic therapies become available. Long-term studies will be crucial to understanding the potential risks and side effects of these interventions.

c. Ethical Use of Enhancement Technologies:

The ability to enhance cognitive function and emotional regulation through dopamine manipulation raises ethical questions about fairness, personal responsibility, and the potential for coercion. How far should we allow individuals to go in their pursuit of cognitive or emotional enhancement? What are the ethical limits of modifying our brains for personal gain?

Conclusion

The future of dopamine research holds transformative potential, with the promise of more effective treatments for mental health conditions, neurodegenerative diseases, and addiction. Advances in genetics, pharmacology, and neuroscience will continue to shape how we understand and manipulate dopamine systems to enhance human performance and well-being.

As we move forward, it is essential to approach these innovations with caution, ensuring that ethical considerations, equity, and safety are prioritized. The next frontier of dopamine research is an exciting one, and it holds the potential to revolutionize how we treat brain disorders, improve cognitive function, and enhance emotional well-being for individuals around the world.

Chapter 22: Integrating Dopamine Mastery into Daily Life – Practical Steps for Personal Transformation

Mastering the balance of dopamine receptors—particularly **DRD1** and **DRD2**—is a powerful tool for enhancing cognitive function, emotional well-being, and overall life satisfaction. While the science behind dopamine signaling provides the foundation for understanding how to regulate mood, motivation, and behavior, true transformation occurs when these principles are effectively integrated into everyday life.

In this chapter, we'll explore practical, actionable steps that can help you apply the knowledge of dopamine receptor functioning to optimize your mental and emotional states. This includes creating a personalized plan to balance **DRD1** and **DRD2**, practical techniques for fostering dopamine equilibrium, and real-life success stories of individuals who have mastered their dopamine systems.

1. Developing a Personalized Plan for Dopamine Balance

The first step in mastering dopamine is understanding how **DRD1** and **DRD2** receptors affect your unique brain chemistry. Everyone has a distinct baseline level of dopamine function, and optimizing it requires an individualized approach. To begin:

a. Assess Your Current Dopamine Function:

Take time to reflect on how you currently feel in relation to motivation, emotional regulation, cognitive function, and addiction. Ask yourself:

- Are you highly motivated or often feel a lack of drive?
- How do you handle stress, anxiety, and other emotional challenges?
- Do you feel mentally sharp, or do you struggle with memory, focus, or decision-making?
- Have you struggled with addiction or habit-forming behaviors?

These reflections will give you insight into your dopamine balance. Are you someone who has heightened motivation but struggles with impulsivity (perhaps indicating overactive **DRD1** receptors)? Or do you tend to experience a lack of motivation and pleasure, signaling possible dysfunction in **DRD2** signaling?

b. Identify Lifestyle Changes That Target Dopamine:

- **Physical Activity**: Exercise is a cornerstone for boosting dopamine receptor sensitivity. Activities like resistance training, cardiovascular exercise, or yoga can help optimize dopamine receptor function, enhancing mood, motivation, and cognition.
- **Diet**: Certain foods can influence dopamine production and receptor sensitivity. **Tyrosine-rich foods** (e.g., lean proteins, bananas, and nuts) support dopamine synthesis, while **antioxidant-rich foods** (such as berries and leafy greens) help protect dopamine-producing neurons. Limiting processed sugars and caffeine can prevent over-stimulation of dopamine receptors.
- **Sleep**: Ensure you're getting **7-9 hours of high-quality sleep** per night. Sleep has a profound effect on dopamine receptor density and function. Poor sleep can lead to a downregulation of **DRD2**, impairing motivation and emotional regulation.

c. Build Mental and Emotional Routines:

- **Mindfulness Meditation**: Practicing mindfulness has been shown to reduce over-activation of dopamine pathways associated with stress while enhancing **DRD1** sensitivity. Integrating 10–20 minutes of daily meditation can improve focus, emotional balance, and overall mental clarity.
- **Goal-Setting and Rewarding**: The **DRD1** receptor is closely tied to goal-directed behavior. Establishing clear, achievable goals and rewarding yourself upon completion (whether through tangible rewards or self-praise) can keep **DRD1** activity at optimal levels. These micro-rewards help reinforce positive behaviors, creating a feedback loop that motivates continued progress.

2. Daily Practices to Foster Dopamine Equilibrium

While long-term strategies for dopamine balance are important, incorporating consistent daily habits can make a significant impact on your emotional and cognitive well-being. Below are several practical techniques you can implement to regulate **DRD1** and **DRD2** activity in your everyday life:

a. Embrace the Power of Small Wins:

The **DRD1** receptor is particularly responsive to achievement and reward. To keep your motivation levels high, break down larger tasks into smaller, manageable goals. Celebrate each win—whether it's completing a project or simply making progress on a larger task. This can be a great way to maintain a steady dopamine flow without overwhelming your system with too much stimulation at once.

b. Engage in Creative Activities:

Creative activities, such as writing, drawing, or playing music, stimulate both **DRD1** and **DRD2** receptors. This dual stimulation enhances divergent thinking, problem-solving skills, and overall cognitive flexibility. Even a short burst of creative expression can activate the reward centers of the brain, helping you experience a sense of accomplishment and pleasure.

c. Foster Positive Social Interactions:

Positive, meaningful social connections release dopamine, particularly through the **DRD2** receptor. Make time for activities that involve social engagement, whether it's a meal with friends, a collaborative project at work, or participating in a hobby group. Healthy, supportive relationships are one of the most effective ways to balance dopamine levels, as they encourage both emotional regulation and pleasure.

d. Practice Delayed Gratification:

Chronic instant gratification can lead to dysregulation in dopamine signaling, particularly over-activation of **DRD2** receptors. Practicing delayed gratification by setting boundaries (e.g., avoiding impulsive consumption of social media or junk food) can help improve dopamine regulation. Over time, learning to delay rewards will boost your **DRD1** sensitivity, improving your ability to maintain motivation for long-term goals.

3. Building Resilience Through Dopamine Mastery

Incorporating dopamine management into your daily life also requires resilience. Developing this mental toughness allows you to cope with setbacks and challenges, an area where **dopamine** plays a crucial role. The next step is to work on strengthening your ability to bounce back from difficulties by cultivating emotional intelligence and using dopamine regulation techniques when under stress.

a. Build Coping Mechanisms for Stress:

Stress triggers the **DRD2** receptor's response to reward and motivation, often leading to the depletion of dopamine. To counteract this, it's important to have coping mechanisms in place:

- **Breathing exercises**, such as **box breathing** or **diaphragmatic breathing**, can quickly reduce the impact of stress by lowering cortisol levels and stabilizing dopamine function.
- **Cognitive reframing**, or the practice of viewing challenges as opportunities for growth, can help reduce emotional distress and promote balanced dopamine activity. This technique prevents the brain from overloading on negative feedback loops.

b. Develop Self-Compassion:

A critical component of dopamine balance is self-compassion. When you experience setbacks or failures, **DRD2** dysfunction can cause emotional dysregulation, leading to feelings of shame or frustration. Instead of beating yourself up, practice self-forgiveness and remind yourself that failure is a part of the learning process. **Self-compassion** encourages the **DRD2** receptor to regulate mood more effectively, leading to less emotional volatility.

4. Real-Life Success Stories: Transforming Life with Dopamine Mastery

To better understand the profound impact of dopamine mastery, consider the stories of individuals who have successfully integrated dopamine balance into their daily routines:

Case Study 1: The Executive Who Conquered Burnout

Sarah, a senior executive in a high-pressure industry, was overwhelmed by stress, burnout, and a lack of motivation. Through tailored dietary changes, regular exercise, and the adoption of mindfulness practices, she was able to optimize her dopamine balance. Sarah integrated small wins into her daily routine by breaking down her workload and celebrating progress. As a result, her motivation, focus, and energy levels improved dramatically, and she regained her passion for both her career and personal life.

Case Study 2: The Recovering Addict's Journey

John struggled with addiction to gambling, driven by a dysregulated **DRD2** system that heightened his pleasure-seeking behavior. By adopting a holistic approach that combined therapy, exercise, and a structured goal-setting routine, John was able to restore balance to his dopamine system. Through self-discipline and delayed gratification practices, he slowly regained control of his life, leading to long-term sobriety and a renewed sense of purpose.

Conclusion

Mastering dopamine receptors is not a one-time achievement but an ongoing process of refinement. By integrating the principles of **DRD1** and **DRD2** regulation into daily life, you can unlock the full potential of your mind and emotions. Whether through physical activity, mindfulness, or building positive social relationships, the key to dopamine mastery lies in consistency and self-awareness.

As you begin to apply these practical steps, you will find that optimizing dopamine signaling is not just about improving motivation and mood—it is about transforming the way you experience and navigate life, one step at a time. By taking control of your dopamine balance, you can create lasting personal transformation and thrive in all aspects of life.

Chapter 23: Overcoming Challenges – Common Pitfalls in Dopamine Regulation

Mastering dopamine receptors—particularly **DRD1** and **DRD2**—is a powerful way to enhance cognitive performance, emotional regulation, and overall well-being. However, like any area of self-improvement, achieving optimal dopamine balance is not without its challenges. In this chapter, we will explore common pitfalls in dopamine regulation and provide actionable solutions to overcome them, helping you sustain long-term mental health and emotional stability.

1. The Challenge of Over-Stimulation

One of the most common challenges people face when trying to balance dopamine levels is over-stimulation. With the prevalence of modern technology, constant access to social media, and instant gratification through various digital platforms, the brain can easily become overwhelmed with dopamine-seeking behavior.

Pitfall: Overuse of digital devices, social media, and other sources of immediate pleasure can cause an overactivation of **DRD2**, leading to a decrease in dopamine receptor sensitivity over time. This can result in emotional instability, difficulty focusing, and an increased reliance on external stimuli to feel good.

Solution:

- **Digital Detox**: Set boundaries for screen time and social media use. Aim for **daily or weekly digital detox periods** to reset your brain's dopamine pathways. This allows **DRD2** receptors to regain their sensitivity.
- **Mindful Consumption**: When using digital media, engage with content that brings long-term value, such as educational resources or meaningful connections. Avoid mindlessly scrolling, which promotes temporary satisfaction but overburdens your dopamine system.

2. The Risk of Instant Gratification

The allure of **instant gratification**—whether through sugary foods, quick rewards, or impulsive behaviors—can overwhelm your dopamine system. This constant barrage of easy-to-obtain pleasures can desensitize dopamine receptors, leading to decreased motivation and an inability to appreciate delayed rewards.

Pitfall: The cycle of instant gratification, especially from high-sugar foods or addictive behaviors (e.g., gambling, gaming, shopping), can cause the brain to rely too heavily on **DRD1** activation. Over time, this decreases your ability to experience satisfaction from more meaningful, long-term pursuits.

Solution:

- **Delayed Gratification Practice**: Deliberately choose to delay rewards in your daily life. Start small: wait 10 extra minutes before indulging in a snack or postpone a planned indulgence until after completing a task. This strengthens **DRD1** receptor sensitivity and increases long-term motivation.
- **Reward Systems for Progress**: When working towards a long-term goal, break it down into smaller, manageable tasks. Reward yourself for completing each milestone, which keeps motivation high without relying on instant gratification.

3. Balancing the Dopamine–Reward System

The interaction between **DRD1** and **DRD2** is vital for maintaining a balanced reward system. However, the brain's reward pathways are complex, and when one receptor is overactive or underactive, it can throw off the entire system. An imbalance between **DRD1** and **DRD2** may lead to heightened impulsivity or a lack of motivation, resulting in emotional and cognitive difficulties.

Pitfall: Over-stimulation of **DRD1** may lead to impulsivity, poor decision-making, and a tendency to chase new rewards at the cost of long-term goals. Conversely, **DRD2** dysfunction can cause a lack of pleasure, depression, and decreased motivation.

Solution:

- **Balanced Goal-Setting**: Engage in goal-setting practices that promote both **short-term rewards** (activating **DRD1**) and **long-term rewards** (promoting **DRD2**). By balancing both types of goals, you activate both receptors in a controlled and sustainable way.
- **Behavioral Self-Regulation**: Regularly check in with yourself to assess whether your dopamine system feels balanced. Ask: Are you seeking constant rewards? Are you avoiding tasks that require focus and effort? Being mindful of these tendencies helps you recalibrate your approach and maintain dopamine balance.

4. The Trap of Negative Emotions and Stress

Stress, anxiety, and negative emotions can trigger the release of dopamine in ways that destabilize emotional regulation. **DRD1** and **DRD2** receptors can become dysregulated under chronic stress, leading to emotional volatility, impulsivity, and even addiction as a coping mechanism.

Pitfall: Chronic stress, lack of emotional control, or negative emotions can lead to the overactivation of **DRD1** and **DRD2**, particularly when stress is linked to avoidance behaviors (e.g., substance use, overeating). This creates a vicious cycle where stress leads to more stress and maladaptive behaviors.

Solution:

- **Stress Management**: Implement relaxation techniques such as **deep breathing**, **progressive muscle relaxation**, and **guided visualization** to manage stress. Mindfulness and meditation, which promote both **DRD1** and **DRD2** regulation, are particularly effective in reducing stress levels.
- **Emotional Resilience Training**: Cultivate emotional resilience by practicing **cognitive reframing**—the process of reinterpreting negative thoughts in a more constructive way. This prevents the brain from resorting to dopamine-sapping behaviors in response to emotional discomfort.

5. The Danger of Over-Reliance on Supplements and Stimulants

In the quest to optimize dopamine function, some individuals may turn to **nootropics**, **dopamine-boosting supplements**, or **stimulants** like caffeine to enhance cognitive performance. While these substances can offer short-term boosts, they can also disrupt the delicate balance of dopamine signaling, leading to dependency or further dysregulation.

Pitfall: Over-reliance on stimulants and supplements can cause an artificial increase in dopamine levels, overstimulating **DRD1** and **DRD2** receptors. Over time, this may lead to receptor desensitization, requiring higher doses for the same effect and potentially causing negative side effects such as anxiety, sleep disturbances, and emotional instability.

Solution:

- **Natural Approaches First**: Prioritize lifestyle changes, such as **exercise, nutrition**, and **mindfulness**, before turning to supplements or stimulants. These natural approaches enhance dopamine receptor function without the risk of dependence or overstimulation.
- **Mindful Supplementation**: If you choose to use nootropics or supplements, do so in moderation and under the guidance of a healthcare provider. Avoid daily dependence on substances that artificially boost dopamine levels.

6. The Need for Consistency and Patience

One of the most important aspects of mastering dopamine regulation is **patience**. Dopamine systems take time to adapt to new habits, and rapid changes in behavior can sometimes cause temporary setbacks. Individuals often become frustrated when immediate results are not achieved, leading to discontinuation of dopamine-regulating practices.

Pitfall: Expecting rapid results from lifestyle changes or new techniques can lead to frustration and abandonment of the practices that help balance dopamine levels. This often occurs when individuals fail to see immediate improvements in motivation or mood.

Solution:

- **Focus on Incremental Progress**: Understand that dopamine balance is a long-term journey. Celebrate small wins along the way, and resist the urge to compare your progress to others. Small, consistent changes will lead to greater results over time.
- **Create a Support System**: Surround yourself with individuals who encourage and support your journey toward dopamine mastery. A supportive environment helps maintain motivation during challenging periods.

Conclusion: Achieving Sustainable Dopamine Balance

Mastering dopamine receptors is an ongoing process that requires mindfulness, consistency, and adaptability. While there will always be challenges along the way—whether from external pressures, impulsive behaviors, or emotional turbulence—the key to overcoming these obstacles lies in understanding the science of dopamine regulation and applying it thoughtfully in daily life.

By being aware of the common pitfalls and taking proactive steps to mitigate them, you can achieve a more balanced and rewarding life. Through consistent practice, you will strengthen both **DRD1** and **DRD2** function, allowing you to navigate life with enhanced motivation, emotional clarity, and overall well-being.

Chapter 24: Building a Dopamine-Friendly Future – The Impact on Society and Humanity

As individuals increasingly grasp the importance of mastering dopamine receptors—**DRD1** and **DRD2**—for personal health, it's important to recognize that the power of dopamine regulation extends far beyond individual well-being. The potential benefits of a society that understands and optimizes its dopamine pathways are profound. In this chapter, we will explore the broader implications of dopamine mastery and how it can contribute to a healthier, more productive, and emotionally resilient world.

1. The Societal Shift Toward Well-Being

In a world where stress, mental health issues, and a reliance on external stimulants (such as social media, drugs, and entertainment) are increasingly prevalent, the science of dopamine regulation offers a transformative solution. Mastering **DRD1** and **DRD2** receptors is not just a personal goal; it holds the potential to reshape societal norms, improving collective mental health and fostering a culture of well-being.

Impact:

- **Mental Health**: With the understanding of dopamine's role in mental health, society can shift toward a preventative, rather than reactive, approach to emotional and psychological well-being. By integrating dopamine mastery into education, healthcare, and corporate settings, we can reduce the burden of mental health disorders, particularly those linked to **dopamine dysregulation**, such as depression, anxiety, and addiction.
- **Emotional Resilience**: A society that values emotional regulation and mindfulness would likely experience fewer instances of burnout, chronic stress, and emotional instability. As individuals learn to balance **DRD1** and **DRD2**, they can better navigate life's challenges, promoting stronger relationships and healthier communities.

2. The Role of Dopamine in Education and Productivity

The mastery of dopamine receptors can also have a profound impact on productivity and cognitive development. By understanding how **DRD1** and **DRD2** contribute to motivation, memory, and learning, educational systems and work environments can be optimized for peak performance.

Impact:

- **Education**: Schools and universities can incorporate the science of dopamine into curricula, teaching students how to manage their dopamine levels for sustained focus, motivation, and academic success. Techniques such as delayed gratification, goal-setting, and mindfulness could be used to enhance learning outcomes, preparing students to thrive in an increasingly complex world.
- **Workplaces**: The application of dopamine mastery in the workplace can revolutionize productivity and job satisfaction. By promoting a balanced reward system, businesses can reduce burnout, encourage collaboration, and foster an environment where employees are intrinsically motivated rather than driven by external rewards. This leads to greater job satisfaction, better employee retention, and improved organizational outcomes.

3. Transforming Health Systems with Dopamine Insights

The medical field stands to benefit significantly from a deeper understanding of **DRD1** and **DRD2** receptors. By incorporating dopamine regulation into treatment protocols, healthcare providers can improve patient outcomes in a wide range of mental health, neurological, and behavioral conditions.

Impact:

- **Mental Health Treatment**: Traditional treatments for mental health disorders, such as depression, bipolar disorder, and schizophrenia, often focus on symptom management through medication. A dopamine-informed approach could offer more personalized and effective interventions, targeting the root causes of dopamine dysregulation. Cognitive-behavioral therapies (CBT), mindfulness practices, and neurofeedback techniques could be integrated with pharmacological treatments for optimal results.
- **Addiction Recovery**: Understanding the roles of **DRD1** and **DRD2** in addiction opens up new pathways for recovery. Treatments could focus on restoring dopamine balance, reducing cravings, and improving emotional regulation. More effective rehabilitation strategies would emerge, helping individuals break free from addictive behaviors and reclaim their lives.

4. Fostering Social Connection and Community Building

Human connections are deeply influenced by dopamine release. Positive social interactions, empathy, and community-building efforts all trigger dopamine pathways, which strengthen our emotional bonds. A society that prioritizes dopamine regulation has the potential to create healthier, more supportive communities.

Impact:

- **Building Stronger Relationships**: By understanding how **DRD1** and **DRD2** receptors influence social bonding, we can create more compassionate and connected communities. Social networks that emphasize emotional intelligence, empathy, and healthy dopamine signaling will foster environments where individuals feel seen, heard, and valued.
- **Collective Mental Health**: A culture that nurtures social connection and values emotional well-being can significantly reduce social isolation, which has been linked to numerous mental health challenges. Initiatives such as community-building programs, peer support groups, and collaborative projects can help individuals find meaning and purpose in their relationships, boosting overall life satisfaction.

5. Reducing Inequality and Promoting Equal Access to Well-Being

The science of dopamine regulation can also play a key role in reducing societal inequalities. When individuals understand the connection between lifestyle factors and dopamine receptor function, they can make more informed decisions about how to optimize their health, regardless of their socio-economic status. However, access to information, resources, and healthcare interventions must be equitable for all members of society.

Impact:

- **Equitable Access to Dopamine Health**: Public health campaigns and educational initiatives can empower marginalized communities to take control of their dopamine health. By providing access to mental health resources, nutritional guidance, and physical activity programs, we can level the playing field, ensuring that all individuals have the tools to master their dopamine systems and achieve optimal well-being.
- **Breaking the Cycle of Poverty and Mental Health**: Understanding dopamine regulation may also help break the intergenerational cycle of poverty and mental health issues. By supporting families with resources for emotional regulation, addiction prevention, and stress management, we can improve outcomes for future generations, reducing the long-term societal costs associated with mental health disorders.

6. Ethical Considerations in a Dopamine-Optimized Society

As we look to a future where dopamine mastery becomes more widespread, it's crucial to consider the ethical implications of widespread dopamine manipulation. While optimizing dopamine pathways can lead to positive outcomes, it's important to maintain a balance between personal enhancement and societal responsibility.

Impact:

- **Ethical Use of Dopamine-Enhancing Technologies**: Advances in biohacking, nootropics, and neurofeedback technologies are making it easier to manipulate dopamine pathways. However, it's essential to ensure that these technologies are used ethically and responsibly. Safeguards must be in place to prevent over-reliance on artificial methods of dopamine enhancement, which could lead to dependence, diminished social interactions, or unintended psychological consequences.
- **Promoting Individual Agency**: While dopamine optimization techniques can provide benefits, they should not be used to manipulate or control individuals. Ensuring that people retain agency over their own health decisions is paramount to avoiding societal pressures that may lead to coercive practices or inequality in access to such technologies.

7. The Vision of a Dopamine-Friendly Future

As we move forward, the knowledge of dopamine receptors—**DRD1** and **DRD2**—can serve as a cornerstone of a new societal paradigm focused on well-being, emotional intelligence, and collective growth. By optimizing dopamine balance, we can unlock untapped potential in individuals and communities, fostering a world that values emotional and mental health as much as physical health.

Impact:

- **Cultural Shift**: A dopamine-friendly future would be one where individuals prioritize long-term fulfillment over short-term pleasure, balance personal and social needs, and cultivate resilience in the face of challenges. This shift could transform everything from our education systems to the workplace, healthcare, and community-building efforts.
- **Global Mental Health Movement**: Understanding and optimizing dopamine function could spark a global movement to address mental health issues at their root, moving away from the stigma and treatment of symptoms to a more holistic, preventive approach that enhances human flourishing on a global scale.

Conclusion: Building a World Based on Dopamine Mastery

The potential to master **DRD1** and **DRD2** receptors is far-reaching, touching every aspect of human life—from personal development to societal well-being. By unlocking the power of dopamine, we can create a healthier, more productive, and emotionally balanced world where individuals thrive in their relationships, careers, and communities. As we integrate dopamine mastery into our daily lives, we are not only enhancing our own potential but contributing to a broader cultural shift that values mental health, emotional resilience, and well-being. This vision of a dopamine-friendly future is one that we can all work toward, shaping a world where everyone has the tools to achieve optimal happiness and success.

Chapter 25: Conclusion – The Path to Mastery of DRD1 and DRD2 for a Balanced Life

As we conclude this journey into the fascinating world of dopamine receptors, particularly **DRD1** and **DRD2**, it's clear that understanding and mastering these receptors holds the potential to transform not only individual lives but society as a whole. From cognitive enhancement and emotional regulation to mental health and societal well-being, the implications of dopamine mastery are vast and profound.

1. A Recap of Key Insights

Throughout this book, we have explored how **dopamine**, the brain's key neurotransmitter, plays an integral role in shaping our behavior, mood, motivation, and mental clarity. The **DRD1** and **DRD2** receptors, as the primary mediators of dopamine's effects, have specific, though interconnected, functions. **DRD1** is primarily involved in motivation, reward processing, and cognitive flexibility, while **DRD2** governs pleasure, reinforcement, and habit formation.

- **DRD1** is crucial for motivating goal-directed behaviors, shaping our capacity for learning, memory, and adaptability. It has a profound influence on executive functions and decision-making processes.
- **DRD2** is essential for regulating pleasure and satisfaction, playing a central role in addiction, reinforcement learning, and even motor control, as seen in disorders like Parkinson's disease.

By understanding these receptors at a genetic, molecular, and physiological level, we have gained insight into their roles in mental health, addiction, aging, creativity, and performance. Through this knowledge, we are empowered to regulate and optimize dopamine signaling for enhanced cognitive function, emotional stability, and overall well-being.

2. Dopamine as a Key to Personal and Collective Success

The mastery of **DRD1** and **DRD2** receptors extends far beyond individual achievement. As we've seen in the previous chapters, optimizing dopamine signaling can lead to profound improvements in various aspects of life: cognitive function, emotional regulation, creativity, productivity, and mental health.

When we can balance these receptors effectively, we unlock the potential for **greater mental clarity**, **emotional resilience**, and **sustained motivation**. This isn't just about feeling good—it's about **thriving**, cultivating deep satisfaction from meaningful pursuits, and overcoming obstacles with a clear, focused mind.

On a collective level, the widespread adoption of dopamine mastery can lead to a more **emotionally balanced, motivated, and healthy society**. As we embrace the power of dopamine regulation, we can create environments—whether in education, healthcare, workplaces, or communities—that foster positive growth, social cohesion, and mental well-being for all.

3. A Final Call to Action: Take Control of Your Dopamine Health

Now that we've laid out the science behind dopamine receptors and their impact on various aspects of life, it's time to take action. Mastering **DRD1** and **DRD2** is not a one-time achievement but an ongoing process. It requires consistent attention to **lifestyle choices**, **emotional practices**, and **mental health strategies**. But with the knowledge shared in this book, you are well-equipped to navigate this journey.

Here are the key actions to start your path toward dopamine mastery:

- **Personalize Your Approach**: No two people have the same dopamine profile. Pay attention to your unique responses to dopamine stimuli—whether it's food, social interactions, exercise, or stress—and adjust accordingly.
- **Embrace Mindfulness**: Mindfulness and meditation are powerful tools for regulating dopamine levels. Engage in regular practices to help restore balance and increase your capacity for self-regulation.
- **Optimize Your Environment**: Surround yourself with people and situations that enhance positive dopamine release—such as supportive relationships, stimulating work, and healthy recreational activities.
- **Stay Informed and Adapt**: The science of dopamine is continually evolving. Keep up with new research and be open to adjusting your strategies as you learn more about how dopamine receptors function.

The path to mastering **DRD1** and **DRD2** is one of continuous growth, introspection, and intentionality. It's about creating habits and practices that support your mental, emotional, and physical health. By taking control of your dopamine system, you can unlock your full potential and live a life of deeper meaning, joy, and fulfillment.

4. The Bigger Picture: A Global Shift Toward Mental Well-Being

As more people embrace the principles of dopamine receptor mastery, the collective impact on mental health could be transformative. Imagine a world where:

- **Mental health is prioritized** alongside physical health, with a deeper understanding of the biological mechanisms that underpin emotional regulation and motivation.
- **Communities thrive** in environments that support emotional resilience, healthy dopamine function, and meaningful social connections.
- **Individuals** are empowered to take charge of their own mental health, using tools and strategies based on scientific insights to foster a sense of balance, purpose, and contentment.

By taking responsibility for our dopamine health, we not only enhance our own lives but contribute to a broader societal shift toward a more **emotionally intelligent, resilient, and productive world**. The science of dopamine is not just a tool for self-improvement; it's a pathway to collective flourishing.

5. Looking Toward the Future

The future of dopamine research holds immense promise. As our understanding of the **DRD1** and **DRD2** receptors deepens, we can expect more advanced techniques for optimizing dopamine function—whether through **personalized medicine**, **neurotechnologies**, or **new behavioral practices**. The integration of this knowledge into public health initiatives, educational systems, and workplace environments could lead to a revolution in how we approach mental well-being.

In this future, dopamine will no longer be viewed solely through the lens of addiction or pleasure-seeking. Instead, it will be recognized as a **fundamental driver** of human motivation, learning, and emotional regulation. With this knowledge, we will be better equipped to face the challenges of the modern world and create a society where mental health is not a luxury, but a shared priority.

Conclusion

The mastery of **DRD1** and **DRD2** receptors offers a powerful tool for individuals seeking to enhance their cognitive abilities, emotional regulation, and overall well-being. By applying the science behind these receptors in our daily lives, we can unlock untapped potential and live a life of purpose, balance, and growth. Moreover, by embracing this knowledge on a societal level, we have the opportunity to create a world where emotional resilience, mental clarity, and productive motivation are accessible to everyone.

Take charge of your dopamine health. Start today. The path to a balanced life and a better future begins with you.